Arduino and Raspberry Pi
Sensor Projects
for the Evil Genius™

Arduino and Raspberry Pi Sensor Projects for the Evil Genius™

Robert Chin

New York   Chicago   San Francisco   Athens   London   Madrid
Mexico City   Milan   New Delhi   Singapore   Sydney   Toronto

Library of Congress Control Number: 201794884

McGraw-Hill Education books are available at special quantity discounts to use as premiums and sales promotions or for use in corporate training programs. To contact a representative, please visit the Contact Us page at www.mhprofessional.com.

Arduino and Raspberry Pi Sensor Projects for the Evil Genius™

1 2 3 4 5 6 7 8 9   DSS   22 21 20 19 18 17

ISBN   978-1-260-01089-3
MHID   1-260-01089-9

This book is printed on acid-free paper.

Sponsoring Editor
  Michael McCabe

Editorial Supervisor
  Donna M. Martone

Production Supervisor
  Lynn M. Messina

Acquisitions Coordinator
  Lauren Rogers

Project Manager
  Patricia Wallenburg, TypeWriting

Copy Editor
  James Madru

Proofreader
  Claire Splan

Indexer
  Claire Splan

Art Director, Cover
  Jeff Weeks

Composition
  TypeWriting

About the Author

Robert Chin has a Bachelor of Science degree in computer engineering and is experienced in developing projects on the TI CC3200 SimpleLink, Android, Arduino, Raspberry Pi, and PC Windows platforms using C/C++, Java, Python, Unreal Script, DirectX, OpenGL, and OpenGL ES 2.0. He is the author of *Home Security System DIY PRO Using Android and TI CC3200 SimpleLink*, *Home Security Systems DIY Using Android and Arduino*, and *Beginning Arduino ov7670 Camera Development*. He is also the author of *Beginning Android 3d Game Development* and *Beginning IOS 3d Unreal Games Development*, both published by Apress, and was the technical reviewer for *UDK Game Development*, published by Course Technology CENGAGE Learning. *Beginning Android 3d Game Development* was licensed to Tsinghua University through Tsinghua University Press.

Contents

Introducing the Arduino and Raspberry Pi

THIS CHAPTER WILL INTRODUCE YOU to the Arduino and the Raspberry Pi. First, I give a brief explanation of what the Arduino is. Then I specifically address the Arduino Uno, discussing its general features, including its capabilities and key functional components. This is followed by a discussion of the Arduino Integrated Development Environment (IDE) software, which is needed to develop programs for the Arduino. Each key function of the Arduino IDE is reviewed, followed by a hands-on example giving detailed step-by-step instructions on how to set up the Arduino for development and how to run and modify an example program using the Arduino IDE. Next comes the Raspberry Pi. I discuss what the Raspberry Pi is and the specifications for the Raspberry Pi 3. This section tells you how to set up the Raspberry Pi before using it for the first time. Raspberry Pi hardware features are addressed, including the general purpose input-output (GPiO) pin specifications. The final section includes a hands-on example of how to control a light-emitting diode (LED) using the Raspberry Pi.

What Is an Arduino?

The Arduino is an open-source microcontroller that uses the C and C++ languages to control digital and analog outputs to devices and electronic components and to read in digital and analog inputs from other devices and electronic components for processing. For example, an Arduino can read a signal from a sensor in a home security system that detects the heat that a human being emits. The sensor sends a signal to the Arduino indicating that a person is in the home. After receiving this information, the Arduino can send commands to a camera such as the ArduCAM Mini digital camera to start taking pictures of the intruder. There are many different Arduino models out there. However, to create the examples in this book, you will need an Arduino model with enough pins to connect the components you desire, such as a camera, Bluetooth adapter, and/or motion sensor. Figure 1-1 shows the official Arduino logo.

Note: The official website of the Arduino is www.arduino.cc.

Figure 1-1 Official Arduino logo.

Arduino Uno

There are a great many Arduino products out there, ranging from models that are small and can actually be worn by the user to models with many digital and analog input-output pins. For the projects in this book, I recommend the Arduino Uno, which is an open-source microcontroller that has enough digital ports to accommodate a camera, a Secure Digital (SD) card reader/writer with enough digital and analog ports for other devices, sensors, lights, and any other gadgets that you may require for your own custom projects. The official Arduino Uno board is made by a company called Arduino SRL, formerly Smart Projects, formed by one of the founders of the Arduino (Figure 1-2). The newer official Arduino Uno boards are slightly different in that they are more blue-green instead of blue in color and contain the Genuino logo under the main Arduino logo. The Genuino trademark is used outside the United States as a result of the split between Arduino founders.

There are also unofficial Arduino Uno boards made by other companies. A good way to tell whether a board is official or unofficial is by the color of a component that is located near the Arduino's USB port. This component on an official Arduino board is a metallic gold color. The component on an unofficial board is green. The writing on the components also differs (Figure 1-3).

A number of other companies also manufacture Arduino Uno boards. Because the Arduino is an open-source item, other companies

Figure 1-2   The official Arduino Uno.

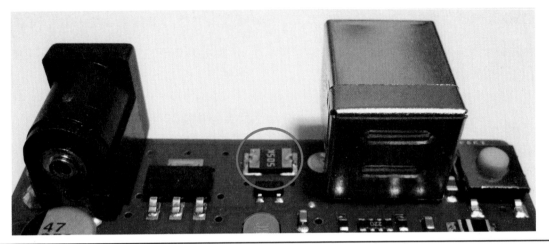

Figure 1-3 Metallic gold colored component on an official Arduino Uno board.

can legally manufacture the board, and the boards vary in quality and price. Generally, an unofficial Arduino Uno costs around $10, and an official Arduino Uno costs around $20. The distinguishing parts of an official Arduino board are the golden component and the high-quality Arduino and Genuino logos printed on the board (Figure 1-4).

Figure 1-4 Key parts of an official Arduino Uno.

Arduino Uno Specifications

- Microcontroller: ATmega328P
- Operating voltage: 5 V
- Input voltage (recommended): 7–12 V
- Input voltage limit: 6–20 V
- Digital I/O pins: 14 (of which 6 provide PWM output)
- PWM digital I/O pins: 6
- Analog input pins: 6
- DC current per I/O pin: 20 mA
- DC current for 3.3-V pin: 50 mA
- Flash memory: 32 kB (ATmega328P) (of which 0.5 kB used by boot loader)
- SRAM: 2 kB (ATmega328P)
- EEPROM: 1 kB (ATmega328P)
- Clock speed: 16 MHz
- Length: 68.6 mm
- Width: 53.4 mm
- Weight: 25 g

Arduino Uno Components

This section covers the functional components of the Arduino Uno.

USB Connection Port

The Arduino Uno has a USB connector that is used to connect the Arduino to the main computer development system via standard USB A male to B male cable so that it can be programmed and debugged (Figure 1-5).

9-V Battery Connector

The Arduino Uno has a 9-V battery connector where you can attach a 9-V battery to power the Arduino (Figure 1-6).

Reset Button

There is a Reset button on the Arduino Uno that you can press to reset the board. This restarts the program contained in the Arduino's memory (Figure 1-7).

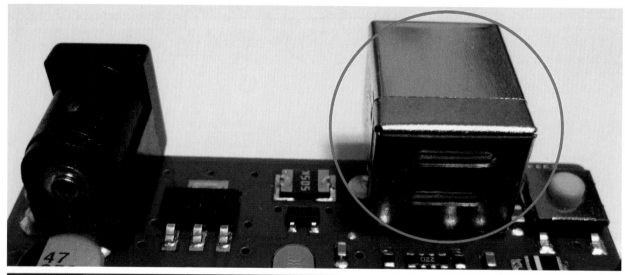

Figure 1-5   USB connector.

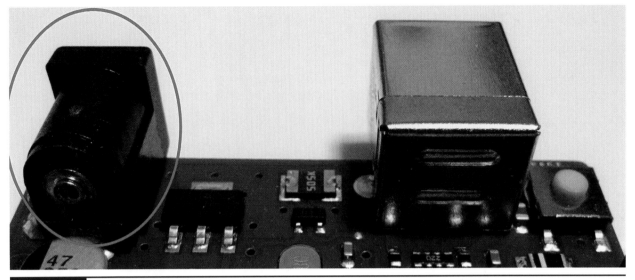

Figure 1-6 Arduino Uno 9-V battery connector.

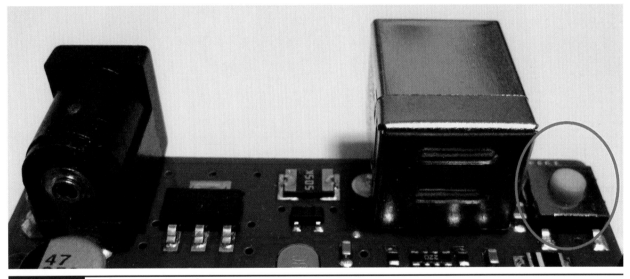

Figure 1-7 Arduino Uno Reset button.

Digital Pins

The Arduino Uno has many digital pins capable of simulating analog output through the process of pulse-width modulation (PWM). For example, a LED light generally has only two modes: on (full brightness) and off (no light emitted). However, with digital PWM, the LED light can appear to have a brightness in between on and off. For instance, with PWM, a LED can start from an off state and slowly brighten until it is at its highest brightness level and then slowly dim until back to the off state. The digital pins on the Arduino Uno are pins 0 through 13. These PWM-capable digital pins are circled in Figure 1-8.

Figure 1-8   Arduino Uno digital pins.

Communication

The communication section of the Arduino Uno contains pins for serial communication between the Arduino and other devices, such as a Bluetooth adapter or a personal computer. The Tx0 and Rx0 pins are connected to the USB port and are used for communication between your Arduino and your computer by means of a USB cable. The Serial Monitor that can be used for sending data to the Arduino and reading data from the Arduino uses the Tx0 and Rx0 pins. Thus you should not connect anything to these pins if you want to use the Serial Monitor to debug your Arduino programs or to receive user input (Figure 1-9). I will talk more about the Serial Monitor later in this book.

The I2C interface consists of an SDA pin (which is pin 4) that is used for data and an SCL pin (which is pin 5) that is used for clocking or driving the device or devices attached to the I2C interface. The SDA and SCL pins are circled in Figure 1-10.

Figure 1-9 Arduino Uno serial communication.

Figure 1-10 Arduino Uno I2C interface.

Analog Input

The Arduino Uno has six analog input pins that can read in a range of values instead of just digital values of 0 or 1. The analog input pin uses a 10-bit analog-to-digital converter to transform voltage input in the range of 0 to 5 V into a number in the range between 0 and 1,023 (Figure 1-11).

Power

The Arduino Uno has outputs for 3.3 and 5 V. One section that provides power is located on the side of the Arduino. You can also provide a separate power source by connecting the positive terminal of the power source to the Vin pin and the ground of the power source to the Arduino's ground. Make sure that the voltage being supplied is within the Arduino's voltage range (Figure 1-12).

The ground connections on the Arduino Uno are shown circled in Figure 1-13.

Figure 1-11 Arduino Uno analog input.

Figure 1-12   Arduino Uno 3.3- and 5-V power outputs.

Figure 1-13   Arduino Uno ground connections.

Arduino Development System Requirements

Arduino projects can be developed on Windows, Mac, and Linux operating systems. The software needed to develop programs that run on the Arduino can be downloaded from the main website at www.arduino.cc/en/Main/Software.

The following is a summary of the different types of Arduino IDE distributions that are available for download. You will only need to download and install one of these files. The file you choose will depend on the operating system your computer is using.

Windows

- **Windows Installer.** This is an .exe file that must be run to install the Arduino IDE.

- **Windows zip file for non-administrator install.** This is a zip file that must be uncompressed in order to install the Arduino IDE. 7-zip is a free file compression and uncompression program available at www.7-zip.org that can be used to uncompress this program.

An Important Note: For Windows XP, I recommend the 1.0.5 r2 version of the Arduino IDE. Later versions may not be stable and may terminate unexpectedly, behave erratically, or may not be able to compile Arduino source code.

Mac

- **Mac OS X 10.7 Lion or newer.** This is a zip file that must be uncompressed and installed for users of the Mac operating system.

Linux

- **Linux 32 bits.** Installation file for the Linux 32-bit operating system.

- **Linux 64 bits.** Installation file for the Linux 64-bit operating system.

The easiest and cheapest way to start Arduino development is probably by using the Windows version on an older operating system such as Windows XP. In fact, the examples in this book were created using the Windows version of the Arduino IDE running on Windows XP. There are in fact many sellers on Ebay from whom you can buy a used Windows XP computer for around $50 to $100. If you are starting from scratch and are looking for an inexpensive development system for the Arduino, consider buying a used computer with Windows XP. The only caution is that support for Windows XP has ended in the United States and some other parts of the world. In China, Windows XP may still be supported with software updates such as security patches.

Arduino IDE Software

Arduino IDE is the program used to develop the program code that runs on and controls the Arduino. For example, to have your Arduino control the lighting state of a LED, you will need to write a computer program in C/C++ using the Arduino IDE. Then you will need to compile the program into a form that the Arduino is able to execute and then transfer the final compiled program using the Arduino IDE. From there the program automatically executes and controls the LED that is connected to the Arduino.

New versions of the Arduino IDE are compiled daily or hourly and are available for download. Older versions of the IDE are also available for downloading at www.arduino.cc/en/Main/OldSoftwareReleases.

This section discusses the key features of the Arduino IDE software. The IDE you are using may be slightly different from the version discussed in this section, but the general functions should still be the same. I won't go in depth into every detail of the IDE because this

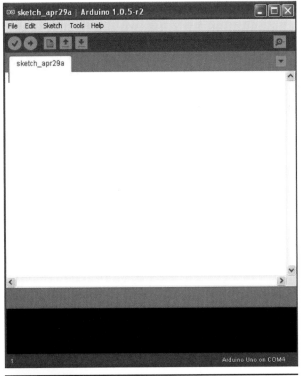

Figure 1-14 The Arduino IDE.

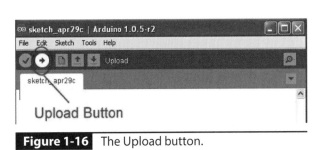

Figure 1-16 The Upload button.

a form the Arduino can execute, and then finally transfers the program via the USB cable connected to your computer to your Arduino board (Figure 1-16).

The New File button creates a new blank file or sketch inside the Arduino IDE, where you can create your own C/C++ program for verification, compilation, and transfer to the Arduino (Figure 1-17).

The Open File button is used to open and load the Arduino C/C++ program source code from a file or load various sample source codes from example Arduino projects that are included with the IDE (Figure 1-18).

section is meant as a quick-start guide and not a reference manual. I will cover the critical features of the Arduino IDE that you will need to get started on the projects in this book (Figure 1-14).

The Verify button checks to see whether the program you have entered into the Arduino IDE is valid and without errors (Figure 1-15). These uncompiled programs are called *sketches*.

The Upload button first verifies that the program in the IDE is a valid C/C++ program with no errors, compiles the program into

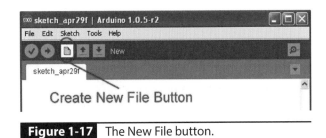

Figure 1-17 The New File button.

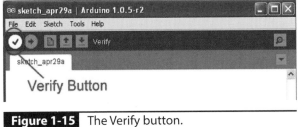

Figure 1-15 The Verify button.

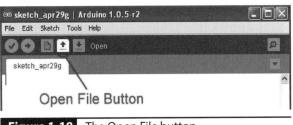

Figure 1-18 The Open File button.

The Save button saves the sketch on which you are currently working to disk. A File Save dialog is brought up first, and then you will be able to save the file on your computer's hard drive (Figure 1-19).

The Serial Monitor button brings up the Serial Monitor debug program, where you can examine the output of debug statements from the Arduino program. The Serial Monitor can also accept user input that can be processed by the Arduino program (Figure 1-20). I will discuss the Serial Monitor and how to use it as a debugger and input console later in this book.

The main window of the Arduino IDE also includes other important features. The title bar of the IDE window contains the Arduino IDE version number. In Figure 1-21, the Arduino version number is 1.0.5 r2. The sketch name is displayed in the source code tab and is "Blink," which is one of the sample sketches that comes with the Arduino IDE. The source code area, which is the large white area with scrollbars on the right side and bottom, is where you enter your C/C++ source code that will control the behavior of the Arduino. The bottom black

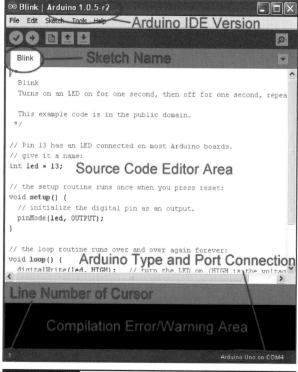

Figure 1-21   The general IDE.

area in the IDE is where warning and errors are displayed from the code verification process. At the bottom left-hand corner of the IDE is a number that represents the line number in the source code where the user's cursor is currently located. In the lower right-hand corner of the IDE is the currently selected Arduino model and COM port to which the Arduino is attached.

Hands-on Example: A Simple Arduino "Hello World" Program with a LED

In this hands-on example, I will show you how to set up the Arduino development system on your Windows-based PC or Mac. First, you need to obtain an Arduino board and USB cable. Then you must install the Arduino IDE and Arduino hardware device drivers. Then I will show you how to load in the "Blink" example sketch. I tell you how to verify that the

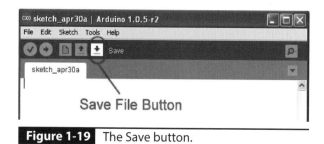

Figure 1-19   The Save button.

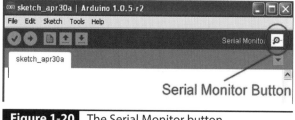

Figure 1-20   The Serial Monitor button.

program is without syntax errors, how to upload it onto the Arduino, and how to tell whether the program is working. Finally, I discuss how the Blink program code works and show you how to modify it.

Get an Arduino Board and USB Cable

You can purchase an official Arduino Uno board from a distributor listed on www.arduino.cc/en/Main/Buy or www.Arduino.org. The first website is still generally considered the main website for Arduino. However, the second website is run by the people actually making Arduino boards. The split among the founders of the Arduino mentioned earlier can be seen here in terms of who is designated as a distributor of an "official" Arduino board.

A second option is to buy an unofficial Arduino Uno made by a seller not listed as an official distributor by either arduino.cc or arduino.org. These boards are generally a lot cheaper than "official" Arduino boards. However, the quality may vary widely between manufacturers or even between production runs of the same manufacturer.

Official Arduino boards generally do not come with a USB cable, but many unofficial boards do come with a short USB cable. Longer Arduino-compatible USB cables (such as 6- or 10-foot cables) can be bought on Amazon.com or eBay.com. I purchased a "Mediabridge USB 2.0 – A Male to B Male Cable (10 Feet) – High-Speed with Gold-Plated Connectors – Black" from Amazon.com for my "official" Arduino Uno, and its seems to work well. Make sure that you get the right kind of USB cable with the right connectors on each end. The rectangular end of the USB cable is connected to your computer, and the square end is connected to your Arduino (Figure 1-22).

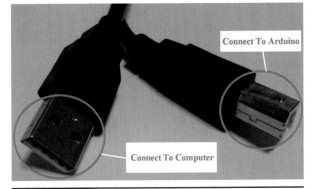

Figure 1-22   Arduino USB cable.

Install the Arduino IDE

The Arduino IDE has versions that can run on Windows, Mac, and Linux operating systems. The Arduino IDE can be downloaded from www.arduino.cc/en/Main/Software. I recommend installing the Windows-executable version if you have a Windows-based computer. Follow the directions in the pop-up windows.

Note: The Arduino website also contains links to instructions for installing the Arduino IDE for Windows, Mac, and Linux at www.arduino.cc/en/Guide/ HomePage. The installation for Linux depends on the exact version of Linux being used.

Install the Arduino Drivers

The next step is to connect your Arduino to your computer using the USB cable. If you are using Windows, it will try to automatically install your new Arduino hardware. Follow the directions in the pop-up windows to install the drivers. Decline to connect to Windows Update to search for drivers. Select "Install the software automatically" as recommended. If you are using XP, ignore the pop-up window warning about the driver not passing Windows logo testing to verify its compatibility with XP.

If this does not work, then instead of selecting "Install the software automatically," specify a specific driver location, which is the "drivers/ FTDI" directory under your main Arduino installation directory.

Load the "Blink" Arduino Sketch Example

Next, we need to load the "Blink" sketch example into the Arduino IDE. Click the Open file button to bring up the menu. Under "01.

Basics," select the "Blink" example to load (Figure 1-23).

The code that is loaded into the Arduino IDE should look like the code in Listing 1-1.

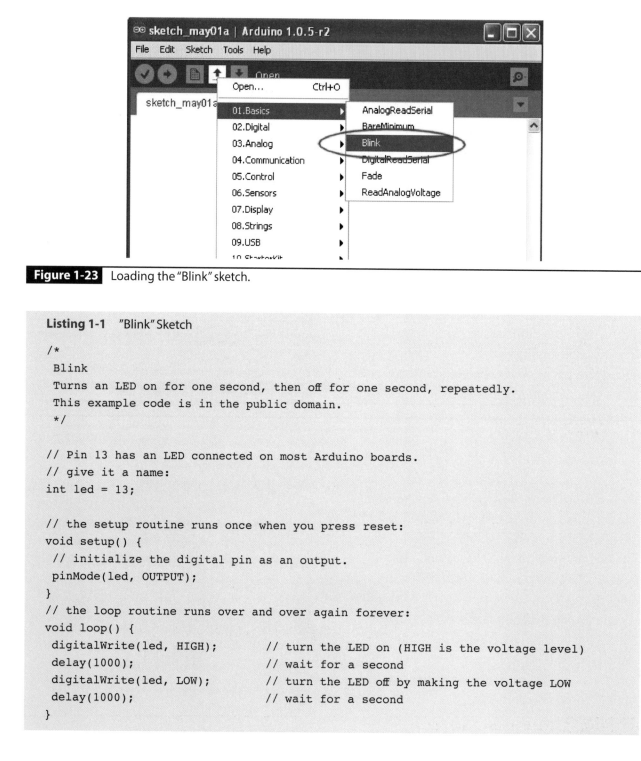

Figure 1-23 Loading the "Blink" sketch.

Listing 1-1 "Blink" Sketch

```
/*
  Blink
  Turns an LED on for one second, then off for one second, repeatedly.
  This example code is in the public domain.
*/

// Pin 13 has an LED connected on most Arduino boards.
// give it a name:
int led = 13;

// the setup routine runs once when you press reset:
void setup() {
  // initialize the digital pin as an output.
  pinMode(led, OUTPUT);
}
// the loop routine runs over and over again forever:
void loop() {
  digitalWrite(led, HIGH);     // turn the LED on (HIGH is the voltage level)
  delay(1000);                 // wait for a second
  digitalWrite(led, LOW);      // turn the LED off by making the voltage LOW
  delay(1000);                 // wait for a second
}
```

Verify the "Blink" Arduino Sketch Example

Click the Verify button to verify that the program is valid C/C++ code and is error-free (Figure 1-24).

Figure 1-24 Verifying the "Blink" sketch.

Upload the "Blink" Arduino Sketch Example

Before uploading the sketch to your Arduino, make sure the type of Arduino under the "Tools > Board" menu item is correct. In our case, the board type should be set to be "Arduino Uno" (Figure 1-25).

Next, make sure that the serial port is set correctly to the one that is being used by your Arduino. Generally, Com1 and Com2 are reserved, and the serial port to which the Arduino will be connected is Com3 or higher (Figure 1-26).

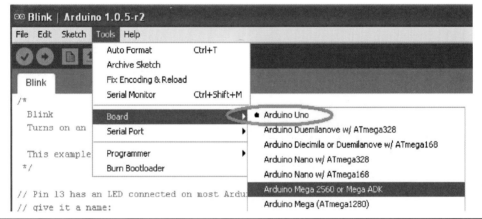

Figure 1-25 Set the Arduino type to "Arduino Uno."

Figure 1-26 Set the serial (Com) port.

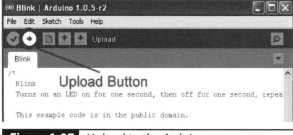

Figure 1-27 Upload to the Arduino.

If you are using a Mac, then the serial port selection should be something like "/dev/tty.usbmodem" instead of a COM*xx* value.

Next, with the Arduino connected, press the Upload button to verify, compile, and then transfer the "Blink" example program to the Arduino. After the program has finished

uploading, you should see a message that the upload has been completed in the Warnings/error window at the bottom of the IDE inside the black window (Figure 1-27).

Note: The Upload button does the job of the Verify button and also uploads the final compiled program to the Arduino.

Final Result

The final result will be a blinking light on the Arduino board near digital pin 13. By design, the Arduino board has a built-in LED connected to pin 13. So this example did not require you to connect an actual separate LED to the Arduino board (Figure 1-28).

Figure 1-28 Built-in LED.

Playing Around with the Code

The default of the program is to turn on the LED for 1 second and then turn it off for 1 second. The code that controls the timing is located in the `loop()` function. The `digitalWrite()` function sets the variable LED, which is pin 13, to either on, which is `HIGH`, or off, which is `LOW`. The `delay()` function suspends execution of the program for 1,000 milliseconds, or 1 second, so that the LED is set on for 1 second and off for 1 second (Listing 1-2).

Play around with the values in the `delay()` functions, lengthening or shortening the time the LED stays on and/or lengthening or shortening the time the LED stays off. For example, to have the LED flash briefly, shorten the first delay value to 100. This will shorten the time that the LED stays on. Upload the new sketch to the Arduino by pressing the Upload button. After it has finished uploading, you will see a message in the black Warnings/error message window at the bottom of the IDE. Look at the LED on the Arduino. The timing of the LED on/off pattern should have changed.

What Is Raspberry Pi?

Raspberry Pi is a small, inexpensive computer based on the Linux operating system. The computer itself sells for around $35 and was originally designed as a learning tool to get young people and students interested in pursuing a career in computer programming. The official website for the Raspberry Pi is located at www.raspberrypi.org.

Raspberry Pi 3 Model B

Many models of the Raspberry Pi have been produced, but the type of Raspberry Pi that I will use in this book is the newest and most popular Raspberry Pi: the Raspberry Pi 3 Model B. This model replaces the Raspberry Pi 2 and is recommended for use in schools and for general purpose use. For those who want to embed a Raspberry Pi as part of a larger project or require a Raspberry Pi that uses low power, then the Raspberry Pi 0 or Model A+ is recommended. The specifications of the Raspberry Pi 3 Model B are as follows:

- 1.2-GHz 64-bit quad-core ARMv8 CPU
- 802.11n wireless LAN
- Bluetooth 4.1
- Bluetooth Low Energy (BLE)
- 1-GB RAM
- Four USB ports
- 40 GPiO pins
- Full HDMI port
- Ethernet port
- Combined 3.5-mm audio jack and composite video
- Camera interface (CSI)
- Display interface (DSI)

Listing 1-2 `loop()` Function

```
void loop() {
  digitalWrite(led, HIGH);        // turn the LED on (HIGH is the voltage level)
  delay(1000);                    // wait for a second
  digitalWrite(led, LOW);         // turn the LED off by making the voltage LOW
  delay(1000);                    // wait for a second
}
```

- Micro SD card slot (now push-pull rather than push-push)
- VideoCore IV 3D graphics core

Raspberry Pi 3 Hardware Overview

The Raspberry Pi 3 is manufactured as a small single board similar to the size of a credit card without a case containing all the above-mentioned components (Figure 1-29).

The Raspberry Pi 3 consists of the following major components, which are labeled in Figure 1-29 as:

- **Item 1:** Two USB ports stacked on top of each other
- **Item 2:** A second set of two USB ports stacked on top of each other
- **Item 3:** Ethernet socket for connecting to a network router

- **Item 4:** A 3.5-mm jack for audio and video output to a monitor or television
- **Item 5:** An MIPi camera serial interface (CSI)
- **Item 6:** A High Definition Multimedia Interface (HDMI) port for video output to a monitor or television
- **Item 7:** GPiO pins that will be used to connect sensor devices to the Raspberry Pi in this book
- **Item 8:** A Display Serial Interface (DSI) that is used to connect the Raspberry Pi to a liquid-crystal display (LCD)
- **Item 9:** A micro USB socket used for power input to the Raspberry Pi

The bottom of the Raspberry Pi 3 contains a micro SD card reader/writer that serves as a storage device like a hard drive would on a desktop computer. The micro SD stores the Raspberry Pi's operating system, program

Figure 1-29 Top view of the Raspberry Pi 3 hardware layout.

Figure 1-30 Bottom view of the Raspberry Pi 3.

applications, and data generated by the program applications and the user. On the Raspberry Pi 3, the micro SD card reader/writer is of the push-pull type. This means that you will need to push the SD card into the slot to have the SD card be read from and written to, and then you will need to pull the card out to remove it from the SD card slot (Figure 1-30).

With the Raspberry Pi oriented like the one in Figure 1-29, the GPiO pin specifications are as shown in Figure 1-31.

The symbols on the pins denote the following:

- **5V:** This is a 5-V output source from the Raspberry Pi that can be used to power your custom circuit.

- **3.3V:** This is a 3.3-V output source from the Raspberry Pi that can be used to power your custom circuit.

- **G:** Ground pin.

- **EE:** EEPROM-related pin (which will not be used in this book).

- **Number:** This is the number of the GPiO pin.

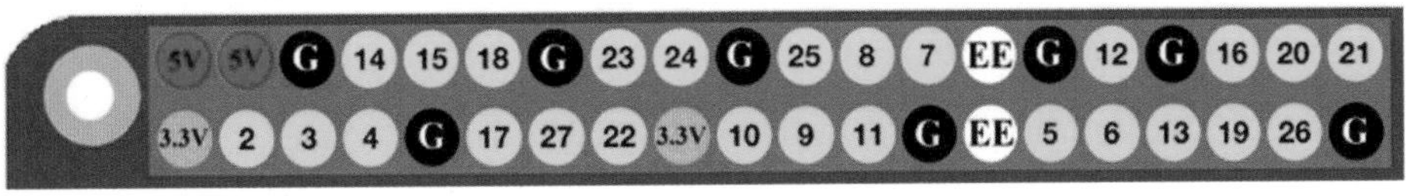

Figure 1-31 Raspberry Pi 2 and Raspberry Pi 3 GPiO pin numbering.

Raspberry Pi System Setup

This section will tell you what you will need to set up your Raspberry Pi computer system and how to perform the actual setup. At the bare minimum, you will need

- A Raspberry Pi 3 board
- A USB keyboard
- A USB mouse
- A micro USB power adapter that can provide 2.5-A power
- An HDMI-compatible monitor or TV (which I recommend) or a monitor with composite inputs or a VGA monitor and an HDMI-to-VGA adapter (I recommend an HDMI monitor because it provides sharp resolution and sound output. The composite output is of a lower-quality resolution, and using a VGA monitor with a Raspberry Pi requires you to make some changes to the Raspberry Pi's configuration file.)
- A micro SD card with new out of box software (NOOBS) on it
- Preferably some kind of case for your Raspberry Pi 3 board to protect it against static electricity and small spills of liquid

If you are just beginning in Raspberry Pi development, then I recommend that you get one of the Raspberry Pi 3 starter kits, which are widely available online on such websites as Amazon.com. Many of the starter kits have all of the items listed above. In addition, if you do not have an HDMI monitor, you can usually find a small one for around US$50 on Amazon.com. These monitors are generally around 10 inches in display size and can be used as portable monitors for your Raspberry Pi 3 system.

To start using your Raspberry Pi, you should

1. Connect the keyboard, mouse, and HDMI monitor to your Raspberry Pi.

2. Insert your SD card with the NOOBS into the Raspberry Pi's micro SD card reader/writer.

3. Turn on your Raspberry Pi by plugging in the micro USB power adapter. An important thing to note is that the Raspberry Pi does not have an on/off switch.

4. The NOOBS setup program will start running. This program will install the actual Raspberry Pi operating system, which is called *Raspbian*. Follow the directions that appear on the screen to install the software. You should see a pop-up window that looks like the one shown in Figure 1-32. Check

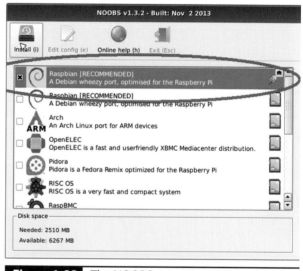

 Figure 1-32 The NOOBS operating system installation screen.

the box next to the Raspbian selection, and click on the Install button.

Python IDLE Development Environment

The programming language that we will be using in this book for projects involving the Raspberry Pi is the Python programming language. The Raspberry Pi has a Python development environment called *IDLE*. IDLE provides you with a read-evaluate-print-loop (REPL) that allows you to continuously enter Python statements and have them executed by the Python language. The output of these statements is also displayed within IDLE. You can start the Python IDLE program by clicking on "Python 3" under the Menu > Programming menu (Figure 1-33).

From within the IDLE program you can also create and save new files. An alternative to using IDLE to create Python programs is to use a

Figure 1-33 The Python 3 development environment.

text editor such as Vim, Nano, or LeafPad and execute the Python program by typing "python filename.py" on the command line.

Hands-on Example: A Simple Raspberry Pi "Hello World" Program with a LED

This section presents a simple hands-on example of creating a blinking LED using the Raspberry Pi 3.

Parts List

For this hands-on example, you will need

- One LED
- One breadboard
- Two wires with one end having a female plug to connect to the Raspberry Pi's GPiO pins and the other end having a male plug to connect to the breadboard

Setting Up the Hardware

1. Insert the short leg of the LED, which is the negative side, into a hole on the breadboard.

2. Insert the long leg of the LED, which is the positive side, into a different hole on the breadboard that is disconnected from the node in step 1. *Note:* Breadboards generally have horizontal rows of nodes of about five holes each that are connected together. Some breadboards have a positive row of holes that are connected together and a negative row of holes that are connected together.

3. Attach a wire from a ground pin on the Raspberry Pi to the ground node on the

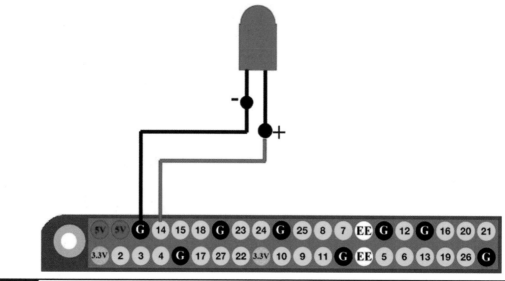

Figure 1-34 Blinking LED on a Raspberry Pi 3 board.

breadboard that contains the negative side of the LED.

4. Attach a wire from GPiO pin number 14 to the node on the breadboard that contains the positive side of the LED.

Figure 1-34 shows a simplified graphic representation of how to hook the LED up to the Raspberry Pi 3 board.

Setting Up the Software

Now you will need to set up the software that will run on the Raspberry Pi itself to control the hardware you just set up. In order to control the blinking of a LED from the Raspberry Pi, you need to do the following:

1. Import the LED class from the `gpiozero` library.

2. Import the sleep function from the `time` library.

3. Create a new LED object that is connected to GPiO pin 14, and then assign this object to the `led` variable.

4. Execute the following in an infinite loop:

   a. Turn the LED on by calling the `led.on()` function.

   b. Wait for 1 second by calling the `sleep()` function with 1 as a parameter.

   c. Turn off the LED by calling the `led.off()` function.

   d. Wait for 1 second by calling the `sleep()` function with 1 as a parameter.

See Listing 1-3.

Listing 1-3 Blinking LED

```
from gpiozero import LED
from time import sleep
led = LED(14)
while True:
 led.on()
 sleep(1)
 led.off()
 sleep(1)
```

Figure 1-35 The final blinking LED project.

After you finish typing in this program or downloading it from the publisher's website, run the program by typing the command "python filename.py." Filename.py is the filename of the program. The LED should turn on for 1 second and then turn off for 1 second (Figure 1-35).

Play around with the times that the LED turns on and off by changing the parameters to the `sleep()` function.

Summary

In this chapter I introduced you to the Arduino and the Raspberry Pi. I started off by discussing the Arduino Uno and covering the details of the UNO's hardware configuration. Then I covered the Arduino IDE, which is needed to create programs for the Arduino platform. This was followed by a simple hands-on example using the Arduino to control the built-in LED. Next, I covered the Raspberry Pi, starting with the specifications for the Raspberry Pi 3 and concluding with a detailed hardware overview, including the exact specifications for the GPiO pins. Then I showed you how to set up the Raspberry Pi and install its operating system using NOOBS. The chapter ended with another hands-on example, this time using the Raspberry Pi to blink a LED on and off.

## CHAPTER 2

Arduino and Raspberry Pi Programming Language Basics

THIS CHAPTER COVERS the basics of the C/C++ programming language used on the Arduino and the Python programming language used on the Raspberry Pi. You will learn the basic elements of each of the languages that you will need to create programs that control the Arduino board and the Raspberry Pi computer. Various key elements such as data types, constants, and control loops are covered.

Overview of the C/C++ Language for the Arduino

The Arduino uses C and C++ in its programs, which are called *sketches*. This section briefly summarizes key language elements. This is not meant as a reference guide, and ideally, you should have some experience with a programming language similar to C and/or C++.

Comments

- **//**. This signifies a single-line comment that is used by the programmer to document the code. These comments are not executed by the Arduino.
- **/* */**. These enclose a multiline comment that is used by the programmer to document the code. These comments are not executed by the Arduino.

Data Types

- **void.** This type being used with a function indicates that the function will not return any value to the function caller. For example, the setup() function that is part of the standard Arduino code framework has a return type of void.

```
void setup()
{
    // Initialize the Arduino, camera,
    // and SD card here
}
```

- **boolean.** A boolean variable can hold either the value of true or false and is 1 byte in length. For example, in the following code, the variable result is declared of type boolean and is initialized to false:

```
boolean result = false;
```

- **char.** The char variable type can store character values and is 1 byte in length. The following code declares that tempchar is of type char and is an array with 50 elements:

```
char tempchar[50];
```

- **unsigned char.** The unsigned char data type holds 1 byte of information in the range of 0 through 255.

- **byte.** The `byte` data type is the same as the `unsigned char` data type. The following code declares a variable called `data` of type `byte` that is initialized to 0:

```
byte data = 0;
```

- **int.** The `int` data type holds a 2-byte number in the range of −32,768 to 32,767.

- **unsigned int.** This data type is 2 bytes in length and holds a value from 0 to 65,535.

- **word.** This data type is the same as the `unsigned int` type.

- **long.** This data type is 4 bytes in length and holds a value from −2,147,483,648 to 2,147,483,647.

- **unsigned long.** This data type is 4 bytes in length and holds a value between 0 to 4,294,967,295.

- **float.** This is a floating-point number that is 4 bytes in length and holds a value between −3.4028235E+38 to 3.4028235E+38.

- **double.** On the current Arduino implementation, `double` is the same as float with no gain in precision.

- **String.** This is a class object that allows the user to easily manipulate groups of characters. In the following code, a new variable called `Command` of type `String` is declared and initialized to the QQVGA resolution:

```
String Command = "QQVGA";
```

- **array.** An array is a continuous collection of data that can be accessed by an index number. Arrays are 0 based, so the first element in the array has an index of 0. Common types of arrays are character arrays and integer arrays. The following code declares the variable `Entries` as an `array` of type `String` that contains 10 elements. The function `ProcessRawCommandElement()` is then called with element number 2 in the `Entries` array, which is the third element in the array. Remember that 0 is the first element in the array.

```
String Entries[10];
boolean success =
    ProcessRawCommandElement(Entries[2]);
```

Constants

- **INPUT.** This is an Arduino pin configuration that sets the pin as an input pin that allows you to easily read the voltage value at that pin with respect to ground on the Arduino. For example, the following code sets the pin VSYNC on the Arduino as an `INPUT` pin, which allows you to read the voltage value of the pin. The function `pinMode()` is an Arduino function included in the built-in library.

```
pinMode(VSYNC, INPUT);
```

- **OUTPUT.** This is an Arduino pin configuration that sets the pin as an output pin that allows you to drive other electronics components such as a light-emitting diode (LED) or to provide digital input to other devices in terms of `HIGH` or `LOW` voltages. In the following code, the pin WEN is set to OUTPUT using the built-in `pinMode()` function:

```
pinMode(WEN , OUTPUT);
```

- **HIGH** (pin declared as `INPUT`). If a pin on the Arduino is declared as an `INPUT`, then when the `digitalRead()` function is called to read the value at that pin, a `HIGH` value would indicate a value of 3 volts or more at that pin.

- **HIGH** (pin declared as `OUTPUT`). If a pin on the Arduino is declared as an `OUTPUT`, then when the pin is set to `HIGH` with the `digitalWrite()` function, the pin's value is 5 volts.

- **LOW** (pin declared as INPUT). If a pin on the Arduino is declared as an INPUT, then when the digitalRead() function is called to read the value at that pin, a LOW value would indicate a value of 2 volts or less.

- **LOW** (pin declared as OUTPUT). If a pin on the Arduino is declared as an OUTPUT, then when the digitalWrite() function is called to set the pin to LOW, the voltage value at that pin would be set to 0 volts.

- **true.** True is defined as any nonzero number such as 1, −1, 200, 5, etc.

- **false.** False is defined as 0.

The define Statement

The define statement assigns a name to a constant value. During the compilation process, the compiler will replace the constant name with the constant value:

```
#define constantName value
```

The following code defines the software serial data receive pin as pin 6 on the Arduino and the software serial data transmit pin as pin 7 on the Arduino:

```
#define RxD 6
#define TxD 7
```

These definitions are used in defining which pins are to receive and transmit data via the software serial method, which is initialized as follows and is used to communicate with a Bluetooth adapter:

```
SoftwareSerial BT(RxD,TxD);
```

The include Statement

The #include statement brings in code from outside files and "includes" it in your Arduino sketch. Generally, a header or .h file is included that allows access to the functions and classes

inside that file. For example, we can include in a program a Wire.h file, which lets us use the Wire library. The Wire library has functions to initialize, to read data from, and to write data to a device connected to the I2C interface. We need the Wire library to use a device that uses the I2C bus.

```
#include <Wire.h>
```

The Semicolon

Each statement in C/C++ needs to end with a semicolon. For example, when declaring and initializing a variable, you will need a semicolon:

```
const int chipSelect = 48;
```

When you use a library that you included with the #include statement, you will need a semicolon at the end when you call a function:

```
Wire.begin();
```

Curly Braces

Curly braces such as { and } specify blocks of code and must occur in pairs. That is, for every opening brace, there must be a closing brace to match. A function requires curly braces to denote the beginning and end of the function:

```
void Function1()
{
  // Body of Function
}
```

Program loops such as the for statement may also need curly braces:

```
for (int I = 0; I < 9; I++)
{
  // Body of loop
}
```

It is also good practice to use braces in control structures such as the if statement:

```
if (I < 0)
{
  // Body of If statement
}
```

Arithmetic Operators

- **=.** The equals sign is the assignment operator used to set a variable to a value. For example, the following code sets the value of the variable `Data` to the result from the function `CreatePhotoInfo()`:

  ```
  String Data = CreatePhotoInfo();
  ```

- **+.** The plus sign performs addition between numbers. It can also be used in other contexts depending on how the addition operator is defined. For example, the following code adds the strings `Command` `PhotoTakenCount` and `Ext` together to get a final string called `Filename`:

  ```
  String    Filename    =    Command    +
     PhotoTakenCount + Ext;
  ```

- **-.** The minus sign performs subtraction. For example, the following code calculates the time it takes to capture a photo using the camera by measuring the difference between the starting time before the image is captured and the ending time just after the image is captured:

  ```
  ElapsedTime   =   TimeForCaptureEnd   -
     TimeForCaptureStart;
  ```

- ***.** The asterisk sign performs multiplication. For example, the total number of bytes in an image is calculated by multiplying the width of the image by the height of the image by the bytes per pixel in the image:

  ```
  int    TotalBytes    =    ImageWidth    *
     ImageHeight * BytesPerPixel;
  ```

- **/.** The back slash sign performs division. For example, the speed in miles per hour of an object is calculated by dividing the number of miles the object has traveled by the number of hours that it took to travel that distance:

  ```
  float Speed = NumberMiles / NumberHours;
  ```

- **%.** The percent sign is the modulo operator that returns the remainder from a division between two integers. For example:

  ```
  int remainder = dividend % divisor;
  ```

Comparison Operators

- **==.** The double equals sign is a comparison operator to test whether the argument on the left side of the double equals sign is equal to the argument on the right side. If the arguments are equal, then it evaluates to true. Otherwise, it evaluates to false. For example, if `Command` is equal to `SystemStart`, then the code block is executed:

  ```
  if (Command == "SystemStart")
  {
      // Execute code
  }
  ```

- **!=.** The exclamation point followed by an equals sign is the not-equal-to operator that evaluates to true if the argument on the left is not equal to the argument on the right side. Otherwise, it evaluates to false. For example, in the following code, if the current camera resolution is not set to VGA, then the code block is executed:

  ```
  if (Resolution != VGA)
  {
      // If current resolution is not VGA
      // then set camera for VGA
  }
  ```

- **<.** The less than operator evaluates to true if the argument on the left is less than the argument on the right. For example, in the following code, the `for` loop will execute the code block while the height is less than the height of the photo. When the height counter becomes equal to or greater than the photo's height, then the loop exits:

  ```
  for (int height = 0; height <
   PHOTO_HEIGHT; height++)
  {
      // Process every row of the photo
  }
  ```

- **>.** The greater than operator evaluates to true if the argument on the left side is greater than the argument on the right side. For example, in the following code, if characters are available characters to read from the Serial Monitor, then the code block executes. That is, the number of available characters to read must be greater than 0.

```
if (Serial.available() > 0)
{
    // Process available characters
    // from Serial port
}
```

- **<=.** The less than sign followed by an equals sign returns true if the argument on the left side is less than or equal to the argument on the right side. It returns false otherwise.

- **>=.** The greater than sign followed by an equals sign returns true if the argument on the left side is greater than or equal to the argument on the right side. It returns false otherwise.

Boolean Operators

- **&&.** This is the AND Boolean operator that only returns true if both the arguments on the left and right sides evaluate to true. It returns false otherwise. For example, in the following code, only if the previous value is O and the current value is K will the code block be executed. Otherwise, it will not be executed.

```
char out,outprev = '$';
if ((outprev == 'O')&&(out == 'K'))
{
    out = ReadinData();
    // Code block
    outprev = out;
}
```

- **||.** This is the OR operator and returns true if either the left side argument or the right side argument evaluates to true. Otherwise, it returns false. For example, in the following

code, if the camera's Command is set to either QQVGA or QVGA, then the code block is executed. Otherwise, it is not executed.

```
if ((Command == "QQVGA") ||
  (Command == "QVGA"))
{
    // Code
}
```

- **!.** The NOT operator returns the opposite Boolean value. The not value of true is false, which is 0, and the not value of false is true, which is nonzero. In the following code, a file is opened on the SD card, and a pointer to the file is returned. If the pointer to the file is NULL, which has a 0 value, then not NULL would be 1, which is true. The if statement is executed when the argument is evaluated to true, which means that the file pointer is NULL. This means that the Open operation has failed, and an error message needs to be displayed.

```
// Open File
InfoFile = SD.open(Filename.c_str(),
  FILE_WRITE);

// Test if file actually open
if (!InfoFile)
{
    Serial.println(F("\nCritical
    ERROR ...
    Can not open Photo Info File for
    output ... "));
    return;
}
```

Bitwise Operators

- **&.** This is the bitwise AND operator between two numbers, where each bit of each number has the AND operation performed on it to produce the result in the final number. The resulting bit is 1 only if both bits in each number are 1. Otherwise, the resulting bit is 0.

- **|.** This is the bitwise OR operator between two numbers, where each bit of each number has the OR operation performed on it to produce the result in the final number. The resulting bit is 1 if the bit in either number is 1. Otherwise, the resulting bit is 0.

- ^. This is the bitwise XOR operator between two numbers, where each bit of each number has the exclusive OR operation performed on it to produce the result in the final number. The resulting bit is 1 if the bits in each number are different and 0 otherwise.

- ~. This is the bitwise NOT operator, where each bit in the number following the NOT symbol is inverted. The resulting bit is 1 if the initial bit was 0 and is 0 if the initial bit was 1.

- <<. This is the "bitshift left" operator, where each bit in the left operand is shifted to the left by the number of positions indicated by the right operand. For example, in the following code, a 1 is shifted left `PinPosition` times, and the final value is assigned to the variable `ByteValue`:

```
ByteValue = 1 << PinPosition;
```

- >>. This is the "bitshift right" operator, where each bit in the left operand is shifted to the right by the number of positions indicated by the right operand. For example, in the following code, bits in the number 255 are shifted to the right `PinPosition` times, and the final value is assigned to the variable `ByteValue`:

```
ByteValue = 255 >> PinPosition;
```

Compound Operators

- ++. This is the increment operator. The exact behavior of this operator also depends on whether it is placed before or after the variable being incremented. In the following code, the variable `PhotoTakenCount` is incremented by 1:

```
PhotoTakenCount++;
```

If the increment operator is placed after the variable being incremented, then the variable is used first in the expression it is in before being incremented. For example, in the following code, the height variable is used first in the for loop expression before it is incremented. So the first iteration of the `for` loop below would use height = 0. So the first iteration of the for loop below would assign h = 0. The h2 variable would be incremented after being used in the expression and assigned to h.

```
int h2 = 0;
for (int height = 0; height <
 PHOTO_HEIGHT; height++)
{
    // Process row of image
    h = h2++;
}
```

If the increment operator is placed before the variable being incremented, then the variable is incremented first, and then it is used in the expression that it is in. For example, in the following code, the h2 variable is incremented first before it is used in the `for` loop. This means that in the first iteration of the loop, the h variable is 1 instead of 0.fint h = 0;

```
int h2 = 0;
for (int height = 0; height <
 PHOTO_HEIGHT; height++)
{
    // Process row of image
    h = ++h2;
}
```

- --. The decrement operator decrements a variable by 1, and its exact behavior depends on the placement of the operator either before or after the variable being decremented. If the operator is placed before the variable, then the variable is decremented before being used in an expression. If the operator is placed after the variable, then the variable is used in an expression before it is decremented. This follows the same pattern as the increment operator discussed previously.

- **+=.** The compound addition operator adds the right operand to the left operand. This is actually a shorthand version of

```
operand1 = operand1 + operand2;
```

which is the same as the version that uses the compound addition operator:

```
operand1 += operand2;
```

- **-=.** The compound subtraction operator subtracts the operand on the right from the operand on the left. For example, the code for a compound subtraction would be

```
operand1 -= operand2;
```

This is the same as

```
operand1 = operand1 - operand2;
```

- ***=.** The compound multiplication operator multiplies the operand on the right by the operand on the left. The code for this is

```
operand1 *= operand2;
```

This is also equivalent to

```
operand1 = operand1 * operand2;
```

- **/=.** The compound division operator divides the operand on the left by the operand on the right. For example,

```
operand1 /= operand2;
```

This is equivalent to

```
operand1 = operand1 / operand2;
```

- **&=.** The compound bitwise AND operator is equivalent to

```
x = x & y;
```

- **!=.** This compound bitwise operator is equivalent to

```
x = x | y;
```

Pointer Access Operators

- ***.** The dereference operator allows you to access the contents to which a pointer points. For example, the following code declares a variable pdata as a pointer to a byte and creates storage for the data using the new command. The pointer variable pdata is then dereferenced to allow the actual data to which the pointer points to be set to 1:

```
byte *pdata = new byte;
*pdata = 1;
```

- **&.** The address operator creates a pointer to a variable. For example, the following code declares a variable data of type byte and assigns the value of 1 to it. A function called FunctionPointer() is defined that accepts as a parameter a pointer to a byte. In order to use this function with the variable data, we need to call that function with a pointer to the variable data:

```
byte data = 1;
void FunctionPointer(byte *data)
{
    // body of function
}
FunctionPointer(&data);
```

Variable Scope

- **Global variables.** In the Arduino programming environment, global variables are variables that are declared outside any function and before they are used:

```
// VGA Default
int PHOTO_WIDTH = 640;
int PHOTO_HEIGHT = 480;
int PHOTO_BYTES_PER_PIXEL = 2;
```

- **Local variables.** Local variables are declared inside functions or code blocks and are only valid inside that function or code block. For example, in the following function, the variable localnumber is only visible inside the Function1() function:

```
void Function1()
{
    int localnumber = 0;
}
```

Conversion

- **char(x).** This function converts a value x into a char data type and then returns it.

- **byte(x).** This function converts a value x into a byte data type and then returns it.

- **int(x).** This function converts a value x into an integer data type and then returns it.

- **word(x).** This function converts a value x into a word data type and then returns it.

- **word(highbyte,lowbyte).** This function combines two bytes, the high-order byte and the low-order byte, into a single word and then returns it.

- **long(x).** This function converts a value x into a long and then returns it.

- **float(x).** This function converts a value x into a float and then returns it.

Control Structures

- **if (comparison operator).** The if statement is a control statement that tests whether the result of the comparison operator or argument is true. If it is true, then execute the code block. For example, in the following code, the if statement tests to see whether more data from the Bluetooth connection need to be read. If so, then read the data in, and assign it to the RawCommandLine variable:

```
if(BT.available() > 0)
{
    // If command is coming in then
    // read it
    RawCommandLine = GetCommand();
    break;
}
```

- **if (comparison operator) else.** The if else control statement is similar to the if statement except with the addition of the else section, which is executed if the previous if statement evaluates to false and is not executed. For example, in the following code, if the frames per second parameter is set to 30 frames per second, then the SetupCameraFor30FPS() function is called. Otherwise, if the frames per second parameter is set to night mode, then the SetupCameraNightMode() function is called.

```
// Set FPS for Camera
if (FPSParam == "ThirtyFPS")
{
    SetupCameraFor30FPS();
}
else
if (FPSParam == "NightMode")
{
    SetupCameraNightMode();
}
```

- **for (initialization; condition; increment).** The for statement is used to execute a code block usually initializing a counter and then performing actions on a group of objects indexed by that incremented value. For example:

```
for (int i = 0 ; i < NumberElements;
 i++)
{
    // Process Element i Here
}
```

- **while (expression).** The while statement executes a code block repeatedly until the expression evaluates to false. In the following code, the while code block is executed as long as data are available for reading from the file:

```
// read from the file until there's
// nothing else in it:
while (TempFile.available())
{
    Serial.write(TempFile.read());
}
```

- **break.** A break statement is used to exit from a loop such as a while or for loop. In the following code, the while loop causes the code block to be executed forever. If data are available from the Serial Monitor, then they are processed, and then the while loop is exited:

```
while (1)
{
   if (Serial.available() > 0)
   {
   // Process the data
   break;
}
```

- **return (value).** The return statement exits a function. It also may return a value to the calling function:

```
return;
return false;
```

Object-Oriented Programming

The Arduino programming environment also supports the object-oriented programming aspects of C++. An example of Arduino code that uses object-oriented programming is the ArduCAM class that was developed to support the ArduCAM minicamera. A C++ class is composed of data, functions that use that data, and a constructor that is used to create an object of the class. An example of class data is

```
byte sensor_model;
```

An example of a constructor that takes the camera model number and the chip select pin as input parameters is

```
ArduCAM(byte model,int CS);
```

An example of a class function is

```
void start_capture(void);
```

Listing 2-1 shows the full class declaration.

Listing 2-1   The ArduCAM Class

```
class ArduCAM
{
   public:
      ArduCAM();
      ArduCAM(byte model,int CS);
      void InitCAM();

      void CS_HIGH(void);
      void CS_LOW(void);

      void flush_fifo(void);
      void start_capture(void);
      void clear_fifo_flag(void);
      uint8_t read_fifo(void);

      uint8_t read_reg(uint8_t addr);
      void write_reg(uint8_t addr, uint8_t data);

      uint32_t read_fifo_length(void);
      void set_fifo_burst(void);
      void set_bit(uint8_t addr, uint8_t bit);
```

(continued on next page)

Listing 2-1   The ArduCAM Class (continued)

```
    void clear_bit(uint8_t addr, uint8_t bit);
    uint8_t get_bit(uint8_t addr, uint8_t bit);
    void set_mode(uint8_t mode);

    int wrSensorRegs(const struct sensor_reg*);
    int wrSensorRegs8_8(const struct sensor_reg*);
    int wrSensorRegs8_16(const struct sensor_reg*);
    int wrSensorRegs16_8(const struct sensor_reg*);
    int wrSensorRegs16_16(const struct sensor_reg*);

    byte wrSensorReg(int regID, int regDat);
    byte wrSensorReg8_8(int regID, int regDat);
    byte wrSensorReg8_16(int regID, int regDat);
    byte wrSensorReg16_8(int regID, int regDat);
    byte wrSensorReg16_16(int regID, int regDat);

    byte rdSensorReg8_8(uint8_t regID, uint8_t* regDat);
    byte rdSensorReg16_8(uint16_t regID, uint8_t* regDat);
    byte rdSensorReg8_16(uint8_t regID, uint16_t* regDat);
    byte rdSensorReg16_16(uint16_t regID, uint16_t* regDat);

    void OV2640_set_JPEG_size(uint8_t size);
    void OV5642_set_JPEG_size(uint8_t size);
    void set_format(byte fmt);

    int bus_write(int address, int value);
    uint8_t bus_read(int address);
  protected:
    regtype *P_CS;
    regsize B_CS;
    byte m_fmt;
    byte sensor_model;
    byte sensor_addr;
};
```

Overview of Python for Raspberry Pi

Python is the programming language we will be using for our projects on the Raspberry Pi. Python is an *interpreted* language, which means that Python programs are not compiled, but each program line is interpreted and run when the Python program is run. Python is also an *object-oriented* language, which means that it supports classes and inheritance. The official Python website is located at www.python.org.

Comments

A single-line comment starts with the pound sign. For example:

```
# This is a comment in Python on the
# Raspberry PI System
```

Statements and Indentation

Statements in Python are different from statements in C/C++ in that they have no ending character. For example, recall the Python hands-on example from Chapter 1, where the LED was turned on by calling:

```
led.on()
```

Notice that the statement has no character to denote the end of the statement. In terms of indicating a block of code instead of using curly braces such as { and } as in C/C++, Python uses indentation as a method of indicating a block of code. For example, in our hands-on example in Chapter 1, the while statement block is defined by indentation. See Listing 2-2.

Listing 2-2 Indentation Defines Code Blocks

```
while True:
 led.on()
 sleep(1)
 led.off()
 sleep(1)
```

Arthmetic Operators

- **+, −, * , /.** Addition, subtraction, multiplication, and division operators
- **//.** Floor division, which performs a division but discards the fractional part of the resulting number
- **%.** Operator that returns the remainder of the division between the two numbers
- **().** Parentheses, which are used to group arithmetic operations
- ****.** Exponent operator, which calculates the result of the left operand raised to the power of the right operand
- **=.** Assignment operator that assigns the value of the right operand to the left operand.

Variable Types

This section covers the types of variables available in the Python language. One of the key differences between Python and C/C++ that was covered earlier in this chapter is that variable types are not declared but are implicitly assumed based on the context of the variable assignment. This helps to speed software development at the expense of error checking.

- **Strings.** Strings can be specified with single quotes such as 'mystring' or double quotes such as "mystring":

```
Teststring = "mystring"
Teststring2 = 'mystring'
```

- **Integers.** Integers are whole numbers that you can use to describe whole objects such as pens, apples, and trees:

```
numberpens = 10
```

- **Floats.** Floats are fractional numbers that can be used to describe partial objects and percentages:

```
Percentcomplete = 5.6
```

- **Lists.** Lists are groups of items separated by commas and enclosed by brackets:

```
winninglotterynumbers = [67, 54, 22, 88,
99]
```

- **Tuple.** A tuple is a series of items separated by commas:

```
mytuple = 94545, 22222, 'testGoodBye!'
```

- **Sets.** A set is an unordered collection of items with no duplicate entries that are separated by commas and enclosed by curly braces. An important note is that even if you assign an item more than once to a set, the duplicates are automatically removed. Also, mathematical operations such as union, intersection, difference, and symmetric difference are supported with sets. For example:

```
Myfruitbasket   =   {'apple',   "lemon",
    'orange',   "strawberry",   'apple',
 'pear', 'orange', 'banana'}
```

- **Dictionaries.** A dictionary is an unordered set of key and value pairs, with the requirement that the keys are unique within that dictionary. The dictionary is enclosed by curly braces, and each entry in the dictionary is separated by a comma. For each entry, the key value pair is denoted by a colon. For example:

```
TelephoneAddressBook = {'john':
    5673344408, 'sam': 4554324139}
```

Control Flow Statements

This section discusses the different types of control flow statements that are available in the Python programming language.

- **if (expression):.** If the `if` expression evaluates to true, which is nonzero, then the code block that follows the `if` statement is executed. Note that there is a colon immediately after the `if` statement, and the code block following the `if` statement is indented.

```
if x< 0:
    Led1.on()
    Led2.off()
```

- **if (expression): elif (expression):.** If the first expression evaluates to true, which is nonzero, then the code block that follows the `if` statement is executed. Otherwise, if the next expression evaluates to true, then the second code block that follows the `elif` statement is executed. For example, in the following code, if `x` is less than 0, then `Led1` and `Led2` are both turned

on. If `x` is not less than 0, then if `x` is greater than 5, both `Led1` and `Led2` are turned off.

```
if x < 0:
    Led1.on()
    Led2.on()
elif x > 5:
    Led1.off()
    Led2.off()
```

- **for Variable in Sequence:.** The `for` statement in Python can iterate over a sequence of items such as that of a list. The following example creates a list of various colors. The `for` statement then iterates through the list, sending each of the colors to the function:

```
colors = ['cyan', 'red', 'brown']
for colorchoice in colors:
    processcolorchoice(colorchoice)
```

- **for Variable in range():.** The `for` statement can iterate over a continuous progression of numbers by using Python's built-in `range()` function. In the following example, the `for` statement iterates over the values 0 through 4. The variable `i` is assigned values from 0 through 4, and each number is printed out to the screen. The `range()` function provides numbers from 0 to 1 less than the input parameter:

```
for i in range(5):
    print(i)
```

- **while (expression):.** The `while` statement executes the code in the indented code block if the expression is true, which is nonzero. The expression is tested at the beginning of the code block, and if the expression is true, the code block is executed, and at the end of the code block, the expression is tested again. This loop repeats until the expression evaluates to false. In the following code, the variable `b` is set to 0. The `while` loop is executed until `b` is equal to or greater than 10. In this loop, `b` is printed to the screen and then incremented by 1:

```
b = 0
while b < 10:
    print(b)
    b = b + 1
```

■ **pass.** The `pass` statement does nothing and is used when a statement is required but the program does not need to take any action. The following code is an infinite loop that requires the `pass` statement to execute:

```
while True:
    pass
```

Python Functions

Python has functions similar to C/C++ language functions. A *function* in Python is defined by the keyword `def` followed by the name of the function and zero or more input parameters. Input parameters can have default values such as a number or a string. The body of a function is indented. The first line of the function body can contain an optional documentation string that is denoted by triple quotes at the beginning and the end of the comment. Next, the rest of the statements belonging to the function are entered. Finally, a function can return a value by using the `return` keyword followed by the value that is to be returned to the calling statement. Returning a value is optional. See the following function example:

```
def FunctionName(inputparameter,
    defaultinteger=5,
    defaultstring="hello"):
    """ Optional Documentation String """
    Statement1
    Statement2
    StatementN
    return returnvalue  #optional return
                        #value
```

Python Modules

Python functions can be easily reused in many different Python programs by putting these functions in a separate file and importing the code as a module. For example, if the following two functions are saved in a file called `myfunctions.py`, then the module that contains these two functions is called `myfunctions`:

```
def FunctionName1(inputparameter,
    defaultinteger=23, defaultstring="first
    function"):
    """ This is the first function in the
    Python module """
    Statement1
    StatementN
```

```
def FunctionName2(inputparameter,
    defaultinteger=53,
    defaultstring="second function"):
    """ This is the second function in the
    Python module """
    Statement1
    StatementN
```

In order to use these two functions in your Python program, you will need to import them into your program by using the `import` command such as

```
import myfunctions
# remainder of python program
```

Python Class Overview

The Python language is object oriented and contains classes such as the C++ language. The basic Python class consists of:

1. The keyword `class` identifies the start of a Python class followed by the name of the class.

2. An optional string that documents the class can follow.

3. Class variables can be initialized such as `integer`, `float`, or `sequence` variables such as lists. These variables will be shared by all instances of the class and thus have one value for all the instances of that class.

4. The class can have an optional constructor like that in C++ that allows the user to initialize the object. The constructor is the method called `__init__(self, ...)` that is called automatically by Python when a class variable is created.

5. Functions in a class have self as a first parameter. This variable can be used to set the specific instance of a local class variable that belongs to the main class. You can use the self variable to set local class variables such as `floats`, `integers`, and `sequences` such as lists. See Listing 2-3.

In order to create an instance of this Python class, you would use statements such as:

```
a = 11
b = 55
x = MyClass(a, b)
```

The variables a and b are created and assigned values of 11 and 55. Next, a new instance of `MyClass` is created with the values stored in a and b and assigned to the x variable. You can access the class variables as in C++ by referencing them with the dot operator as in:

```
# Access InstanceValue1 and
# print it out to the screen.
print(x.InstanceValue1)
```

To call the class functions, you would do the same thing as in:

```
x.function1(b, a) #Calls the function1()
function with b and a as parameters
```

Listing 2-3   Basic Python Class

```
class MyClass:                           #class name is MyClass
 """A simple example class"""            #document string for the class
 ClassIntegerNumber = 12345              #variable shared by all instances of this
class
 ClassFloatNumber = 3.4                  #variable shared by all instances of this
class
 ClassList = []                          #variable shared by all instances of this
class

 def __init__(self, value1, value2):     #constructor for the class
   self.InstanceValue1 = value1          #variable specific to an instance of this class
   self.InstanceValue2 = value2          #variable specific to an instance of this class
   self.InstanceList = []                #variable specific to an instance of this class

 def function1(self, value1, value2 ):   #class function
   self.InstanceValue3 = value1          #variable specific to an instance of this class
   self.InstanceList.append(value2)      #variable specific to an instance of this class
   return "hello world"                  #function returns a value to the caller
```

Python Class Inheritance

A class in Python can inherit from a base class. In the `class` declaration, the base class name is set in parenthesis:

```
class DerivedClassName(BaseClassName):
  Statement1
  StatementN
```

A Python class can also inherit from multiple base classes. In the `class` declaration, the base class names are separated by commas:

```
class DerivedClassName(Base1, Base2,
    Base3):
  Statement1
  StatementN
```

Summary

This chapter covered the basics of the Arduino and Raspberry Pi programming languages. We explored such basic topics as data types, constants, built-in functions, control loops, and classes. In addition, this chapter was meant as a quick start guide to the basics of the programming languages used on the Arduino and Raspberry Pi platforms, not as a reference manual. Please refer to the official Arduino and Python language reference guides and tutorials for more information.

## CHAPTER 3

Basic Electrical Components

THIS CHAPTER DISCUSSES the basic electrical components common to many Arduino and Raspberry Pi projects. The chapter starts by showing you a basic electric circuit. Then the difference between a series and a parallel connection is illustrated, followed by a definition of a resistor and the introduction of some key concepts such as Ohm's law. Then the chapter shows you how to read the color bands on a resistor to determine the resistor's value. Next, light-emitting diodes (LEDs), the red-green-blue (RGB) LEDs, the Piezo buzzers, potentiometers, push buttons, and breadboards are described. The remainder of this chapter provides hands-on example projects involving the Arduino and Raspberry Pi and using the basic electrical components discussed earlier.

Electronics Basics

The basic electric circuit consists of a voltage source, a switch to activate or deactivate the circuit, and an electrical component that uses the voltage source to operate. The voltage induces an electric current to flow from the positive terminal to the negative terminal of the voltage source through the electrical component (Figure 3-1).

Traditionally, a mechanical switch was used to activate an electric circuit. However, in this book, instead of a mechanical switch, we will

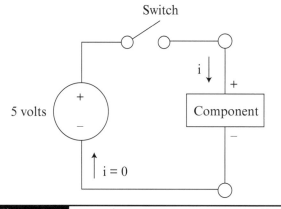

Figure 3-1 Basic electric circuit.

use an Arduino or Raspberry Pi to activate and deactivate the circuit and to provide the voltage to the circuit. A digital pin on the Arduino or Raspberry Pi will turn the circuit on or off by supplying a voltage to a pin. The pin is connected to the positive terminal of the component, and the negative terminal of the component is connected to the ground (also known as negative or common) on the Arduino or Raspberry Pi (Figure 3-2).

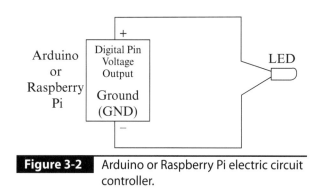

Figure 3-2 Arduino or Raspberry Pi electric circuit controller.

Electrical components can be connected in parallel or series configurations. In a parallel connection, all the positive terminals are connected together, and all the negative terminals are connected together. When a voltage source is connected in parallel with the components, all the components have the same voltage as the voltage source. In Figure 3-3, node A contains all the positive terminals, and node B contains all the negative terminals.

In a series connection between two components, one component's positive terminal is connected to the negative terminal of the other component. In a circuit where the voltage is placed across two components connected in series, the total voltage across both components equals the input voltage. This can be expressed as:

$$V_{source} = v_1 + v_2$$

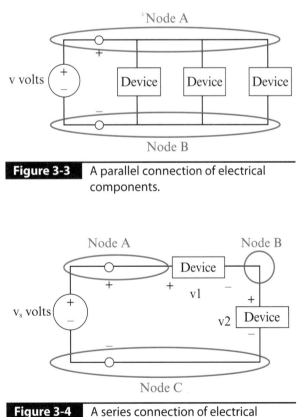

Figure 3-3   A parallel connection of electrical components.

Figure 3-4   A series connection of electrical components.

where V_{source} is the voltage source, v_1 is the voltage across the first component connected in series, and v_2 is the voltage across the second component connected in series with the first component (Figure 3-4).

Resistor

A resistor is a basic electrical component that obeys Ohm's law. The current goes through the resistor from the positive terminal to the negative terminal (Figure 3-5).

Ohm's law can be stated as:

- The voltage across a resistor is equal to the current going through the resistor multiplied by the resistance value, or

- A resistance equals the voltage across the resistor divided by the current going through the resistor, or

- The current going through a resistor equals the voltage across the resistor divided by the resistance (Figure 3-6).

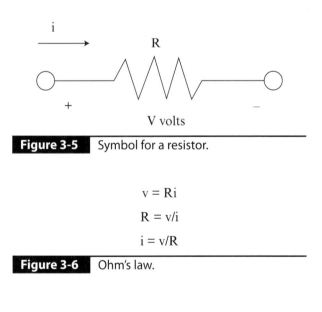

Figure 3-5   Symbol for a resistor.

$$v = Ri$$
$$R = v/i$$
$$i = v/R$$

Figure 3-6   Ohm's law.

Resistor Color Codes

A resistor has a value displayed on the component that is coded with bands of color. The color band closest to the end of the resistor represents the first or leftmost digit, the next band represents the second digit, and the next band represents the number of zeros to add to the first and second digits to create the resistance value. The next color band represents the

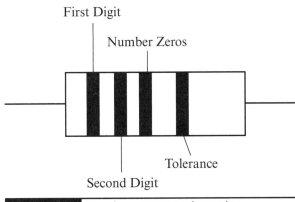

Figure 3-7 Reading resistor color codes.

Color	1st and 2nd Digits	Number Zeros	Tolerance
Black	0	0	—
Brown	1	1	±1%
Red	2	2	±2%
Orange	3	3	—
Yellow	4	4	(±5%)
Green	5	5	±0.5%
Blue	6	6	±0.25%
Violet	7	7	±0.1%
Gray	8	8	±0.05% (±10%)
White	9	9	—
Gold	–	× 0.1	±5%
Silver	–	× 0.01	±10%
None	–	–	±20%

Figure 3-8 Resistor color code chart.

tolerance of the resistor, which is the range of values that the actual resistor could have. If no color band is present, then the tolerance is 20 percent (Figure 3-7). Figure 3-8 illustrates the color banding method.

For example, if a resistor has the color bands of green, brown, red, and yellow, then the first digit of the resistance value is 5, the next digit is 1, the number of zeros after is two, and the tolerance is plus or minus 5 percent. The final value of the resistor thus is 5,100 ohms (Ω) plus or minus 5 percent of that value.

Resistors in Series

If two resistors are connected in series, then the total resistance of the two resistors is the sum of each of the individual resistors, such as:

$$R_{total} = R_1 + R_2$$

More generally speaking, the total resistance of a group of resistors connected in series is

$$R_{total} = R_1 + R_2 + R_3 + R_4 + \cdots + R_N$$

Resistors in Parallel

If two resistors are connected in parallel, then the total resistance of the two resistors is

$$R_{total} = (R_1 * R_2)/(R_1 + R_2)$$

LEDs

The term LED stands for *light-emitting diode*, and LEDs are commonly used as indicators or light sources. A LED has one short leg that is the negative terminal and one longer leg that is the positive terminal. The LED lights up when a voltage is applied across the terminals (Figure 3-9).

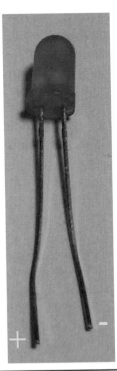

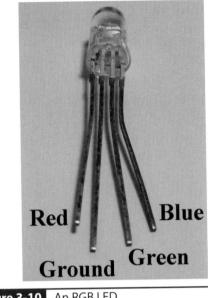

Figure 3-10 An RGB LED.

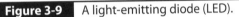

Figure 3-9 A light-emitting diode (LED).

RGB LEDs

Red-green-blue (RGB) LEDs are a type of LED that produces a color based on red, green, and blue inputs to the component. The longest leg of the LED is the negative side or ground terminal. On one side of the ground leg there is another leg that is the red color input to the LED. On the other side are two more legs. The leg closest to the ground leg is the green color input, followed by the leg for the blue color input (Figure 3-10).

Piezo Buzzer

A piezo buzzer has two terminals, one positive and one negative. If you put a voltage across the terminals, the buzzer makes a clicking sound. If you do this many times per second, then you get a tone (Figure 3-11).

Figure 3-11 Top and side views of a piezo buzzer.

Potentiometer

A potentiometer can output a varying voltage based on the position of the knob. This component has a knob that can be used to set the voltage output. On the back of the component, there are three pins. The outermost pins are connected to the positive and negative

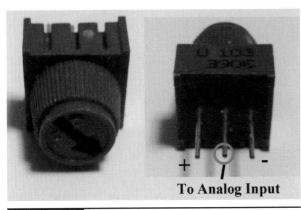

Figure 3-12 A potentiometer.

terminals of a voltage source. The middle pin outputs a voltage value based on the position on the potentiometer's knob and the input voltage source. Output voltage values can range from a maximum voltage value to 0 in smooth analog increments (Figure 3-12).

Push Buttons

Push buttons can be used as user input devices. A button acts like a switch. When the button is not pressed, the two terminals on the button are open and do not conduct electricity. When

Figure 3-13 A push button.

the button is pressed, the two terminals are connected and can conduct electricity. For example, the user can push a button to turn a LED on and off (Figure 3-13).

Breadboards

To build the circuits in this book, you will need a breadboard, which can be used for rapid prototyping of electronics projects without the need for soldering components. A breadboard has rows of about five holes each that are connected together. This node can be used to connect many different items together. On some breadboards, there are also vertical columns of holes that represent either positive or negative terminals of the voltage source (Figure 3-14).

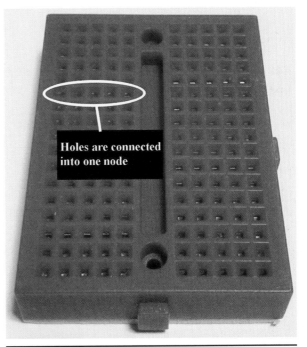

Figure 3-14 A breadboard.

Hands-on Example: Arduino Serial Monitor and LED Indicator

In this hands-on example, you will learn how to turn on and turn off an actual LED using your Arduino. Previously, in Chapter 1, you learned how to use the built-in LED on the Arduino board. You also learned how to use the Serial Monitor to print out debug messages as the LED is turned on and off.

Parts List

The parts you will need for this hands-on example project include

- 2 male-to-male wires
- 1 breadboard
- 1 LED
- (Optional) 1 resistor (10 kΩ recommended) (used to reduce the brightness of the LED)

Setting Up the Hardware

Set up the hardware for this hands-on example by:

- Connecting pin 8 on the Arduino to the positive terminal of the LED.
- Connecting the negative terminal of the LED to one end of a resistor (10 kΩ is suggested). This resistor lowers the brightness of the LED because it lowers the voltage across the LED (see Figure 3-4 for more information about this type of connection which is called a series connection).
- Connecting the other end of the resistor to a ground (GND) pin on the Arduino.

You may need to use a breadboard and groups of nodes on the breadboard to construct the circuit (Figure 3-15).

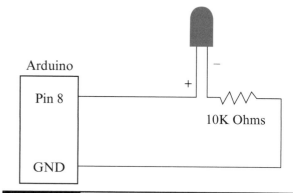

Figure 3-15 Schematic of the project.

Setting Up the Software

Now you need to set up the software on the Arduino that will be used to control the hardware circuit. The Arduino program, or *sketch*, will need to be either downloaded from the publisher's website or typed in. The program does the following:

1. It assigns pin 8 to the variable LED pin that will be used to turn the LED on and off.

2. In the one-time initialization, the function `setup()`:

   a. Sets the pin assigned to the LED as an output pin so that voltage can be provided to that pin by calling the `pinMode()` function.

   b. Initializes the Serial Monitor debug interface by calling the `Serial.begin()` function with 9,600 baud, which sets the baud transfer rate.

   c. Prints out to the Serial Monitor a message that initialization has begun.

3. The main `loop` function, which executes continuously:

   a. Prints out a message to the Serial Monitor that the LED is on.

b. Sets the voltage to the pin connected to the LED to HIGH by calling the `digitalWrite(LEDPin, HIGH)` function with the pin number as the first parameter and the pin state as the second parameter.

c. Suspends execution of the program for half a second by calling the `delay(500)` function with the parameter 500 milliseconds. This leaves the LED light on.

d. Prints out a message to the Serial Monitor that the LED is off.

e. Sets the voltage to the pin connected to the LED to LOW or 0 by calling the `digitalWrite(LEDPin, LOW)` function with the pin number as the first parameter and the voltage state as the second parameter.

f. Suspends execution of the program for half a second by calling the `delay(500)` function with 500 milliseconds as a parameter. This has the effect of leaving the LED off for half a second. See Listing 3-1.

Upload the program to the Arduino, and click on the Serial Monitor button to bring up the Serial Monitor. After a few seconds, you should see a debug message displayed within the Serial Monitor window such as that shown in Figure 3-16.

Listing 3-1 Serial Monitor and LED Blink

```
int LEDPin = 8;

// The setup function runs once when you press reset or power the board
void setup()
{
 // Initialize digital pin as an output.
 pinMode(LEDPin, OUTPUT);
 Serial.begin(9600);
 Serial.println("INITIALIZING Arduino Sketch.........................");
}

// The loop function runs over and over again forever
void loop()
{
 Serial.println("L.E.D. is now High. ");
 digitalWrite(LEDPin, HIGH); // Turn the LED on (HIGH is the voltage level)
 delay(500);

 Serial.println("L.E.D. is now Low. ");
 digitalWrite(LEDPin, LOW); // Turn the LED off by making the voltage LOW
 delay(500);
}
```

```
⊙ COM5

INITIALIZING Arduino Sketch.........................
L.E.D. is now High.
L.E.D. is now Low.
L.E.D. is now High.
L.E.D. is now Low.
L.E.D. is now High.
L.E.D. is now Low.
L.E.D. is now High.
L.E.D. is now Low.
L.E.D. is now High.
L.E.D. is now Low.
L.E.D. is now High.
L.E.D. is now Low.
L.E.D. is now High.
```

Figure 3-16 Debug Serial Monitor output.

You should also see the LED light blink on and off at half-second intervals. The brightness of the LED can be adjusted by raising or lowering the value of the resistor that is attached in series. Try to operate the circuit without the resistor, and you will notice that the LED is much brighter without it.

Hands-on Example: Arduino RGB LED

In this hands-on example, you will learn how to set up and use an RGB LED that can generate red, green, and blue colors.

Parts List

The parts you will need for this hands-on example project include

- 1 RGB LED
- 1 breadboard
- 5 male-to-male wires

Setting Up the Hardware

To set up the hardware for this hands-on example, you will need to:

- Connect a wire from the Arduino's pin 11 to the leg of the LED that handles the blue color input.
- Connect a wire from the Arduino's pin 10 to the leg of the LED that handles the green color input.
- Connect a wire from the Arduino's pin 9 to the leg of the LED that handles the red color input.
- Connect a wire from one of the Arduino's GND pins to the leg of the LED that serves as the GND or common.

Note: You can stick the RGB LED into a breadboard and use the male-to-male wires to connect the Arduino's pins to the correct nodes on the breadboard representing the red, green, blue, and ground terminals (Figure 3-17).

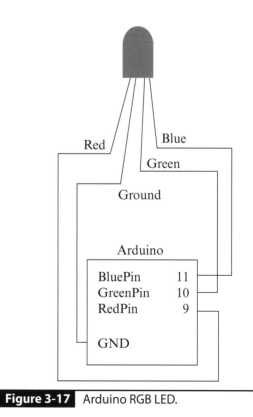

Figure 3-17 Arduino RGB LED.

Setting Up the Software

Now you need to set up the software that controls the RGB LED hardware. For this example, we will use pulse-width modulation (PWM) to simulate the generation of voltage output values between 0 and the maximum output voltage the Arduino supports, which is 5 V. This will give an appearance of a continuous range of analog voltage outputs using a digital method. PWM generates square waves that have values of either 0 or 5 V. The amount of time that the value is at 5 V determines the brightness of the LED. The `analogWrite(PinNumber, Value)` function uses PWM to output square-wave voltage values. The first parameter is the Arduino pin number, and the second parameter is a value between 0 and 255. A value of 0 means that the LED color component is completely off, and a value of 255 means that the LED color component is always on. A value of 127 means that the LED color component is half the maximum brightness. The pins on most Arduinos that support PWM are pins 3, 5, 6, 9, 10, and 11.

The program that controls the RGB LED:

1. Sets the Arduino pins that will output color information to the red, green, and blue inputs on the RGB LED.

2. Uses the `setup()` function to initialize the Serial Monitor, which prints a message indicating that the program has started to run.

3. Uses the `Reset()` function to clear the values being output to all the color component pins to 0, which is completely off.

4. Uses the `FadeLED(int PinNumber, int MinValue, int MaxValue, int Increment, int Delay)` function to increase or decrease the brightness of the RGB LED component using PWM. The first parameter is the pin number on the Arduino to use, and the next two parameters are the minimum and maximum value range to use when fading the LED up or down. The next parameter is the increment by which to increase or decrease the LED brightness, and this value can be either positive or negative. The final parameter is the value in milliseconds between each increase or decrease in the PWM value.

5. The main `loop()` function:

   a. Resets the RGB LED to off by calling the `Reset()` function.

   b. Fades the red component of the RGB LED upward to full brightness and then fades it down until it is off.

   c. Fades the green component of the RGB LED upward to full brightness and then fades it down until it is off.

   d. Fades the blue component of the RGB LED upward to full brightness and then fades it down until it is off. See Listing 3-2.

Listing 3-2   Arduino RGB LED Example

```
int BluePin = 11;
int GreenPin = 10;
int RedPin = 9;

void setup()
{
 Serial.begin(9600);
 Serial.println("INITIALIZING RGBLED Sketch.........................");
}

void Reset()
{
 // Clear All Colors
 analogWrite(RedPin, 0);
 analogWrite(GreenPin, 0);
 analogWrite(BluePin, 0);
}

void FadeLED(int PinNumber, int MinValue, int MaxValue, int Increment, int Delay)
{
 // Test Direction
 if (Increment > 0)
 {
    // Fade Up
    for (int i = MinValue; i <= MaxValue; i=i+Increment)
    {
       analogWrite(PinNumber, i);
       delay(Delay);
    }
 }
 else
 {
    // Fade Down
    for (int i = MaxValue; i >= MinValue; i=i+Increment)
    {
       analogWrite(PinNumber, i);
       delay(Delay);
    }
 }
}
 void loop()
 {
  Reset();
  FadeLED(RedPin, 0, 255, 1, 10);
  FadeLED(RedPin, 0, 255, -1,10);

  FadeLED(GreenPin, 0, 255, 1, 10);
  FadeLED(GreenPin, 0, 255, -1,10);

  FadeLED(BluePin, 0, 255, 1, 10);
  FadeLED(BluePin, 0, 255, -1,10);
  }
```

Figure 3-18 Demonstration of an RGB LED fading up and down in red, green, and blue colors.

When you run the program, you should see something similar to what is shown in Figure 3-18.

Hands-on Example: Arduino LED Buzzer Siren

In this hands-on example, you will learn how to use a buzzer in combination with a LED to produce a flashing red indicator with a buzzer whose pitch increases as the brightness of the LED increases and decreases with decreasing light intensity.

Parts List

The parts you will need for this hands-on example project include

- 1 LED
- 1 buzzer
- 1 breadboard
- 6 wires (approximately)

Setting Up the Hardware

Set up the circuit for this project by:

- Connecting pin 9 on the Arduino to the positive terminal of the LED.
- Connecting pin 7 on the Arduino to the positive terminal of the buzzer.

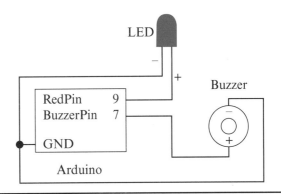

Figure 3-19 Schematic for the LED buzzer siren.

- Connecting the negative terminal of the LED to a GND pin on the Arduino.
- Connecting the negative terminal of the buzzer to a GND pin on the Arduino (Figure 3-19).

Setting Up the Software

Now you need to set up the software that controls the circuit for this example project. To generate sounds with the buzzer, you need to use the built-in `tone()` function. This function, which is defined as `tone(pin, frequency, duration)`, generates a square wave. The first parameter is the pin number on which to generate the square wave. The second parameter is the frequency of this wave, and the third parameter is the optional duration in milliseconds to generate the square wave. The `tone()` function interferes with pulse-width modulation (PWM) on pins 3 and 11 of Arduino boards other than the Arduino Mega, so it is not recommended to use PWM when using these pins on an Arduino Uno.

In the LED buzzer siren program:

1. Pin 9 on the Arduino is set to control the LED.

2. Pin 7 on the Arduino is set to control the buzzer.

3. The setup() function initializes the Serial Monitor and prints out a message indicating that the program has started.

4. The FadeLED() function is defined as FadeLED(int PinNumber, int MinValue, int MaxValue, int Increment, int Delay, int StartFrequency, int FreqIncrement). This function fades the LED up or down as well as increases or decreases the pitch of the tone emitted by the buzzer. The first parameter is the pin number to which the LED is connected. The next two parameters specify the fading range for the LED. The fourth parameter indicates the increment by which to increase or decrease the brightness of the LED. The fifth parameter specifies the delay in milliseconds between increments or decrements in brightness. The sixth parameter is the beginning frequency from which the pitch will either rise or fall. The last parameter is the amount the frequency will change each time the brightness of the LED is changed. This function also returns the ending frequency after the LED's fade-up or fade-down is completed. This is helpful in executing a smooth, continuous tone between one fade-up/down and the next fade-up/down.

5. The loop() function:

    a. Fades the LED up by calling the FadeLED() function. The starting frequency for the buzzer is set to 300. After the fade-up is completed, the final value for the frequency of the buzzer is saved in the EndFreq variable.

    b. Fades the LED down by calling the FadeLED() function. The starting frequency for the buzzer is set to the last frequency generated by the previous FadeLED() function, which was stored in the EndFreq variable. See Listing 3-3.

Listing 3-3 The LED Buzzer Siren Program

```
int RedPin = 9;
int BuzzerPin = 7;

void setup()
{
 Serial.begin(9600);
 Serial.println("INITIALIZING L.E.D. Alarm Sketch.........................");
}

int FadeLED(int PinNumber, int MinValue, int MaxValue, int Increment, int Delay,
 int StartFrequency, int FreqIncrement)
{
 int CurrentFrequency = StartFrequency;

 // Test Direction
 if (Increment > 0)
```

```
{
    // Fade Up
    for (int i = MinValue; i <= MaxValue; i=i+Increment)
    {
        analogWrite(PinNumber, i);
        tone(BuzzerPin,CurrentFrequency);
        delay(Delay);
        CurrentFrequency += FreqIncrement;
    }
}
else
{
    // Fade Down
    for (int i = MaxValue; i >= MinValue; i=i+Increment)
    {
        analogWrite(PinNumber, i);
        tone(BuzzerPin,CurrentFrequency);
        delay(Delay);
        CurrentFrequency -= FreqIncrement;
    }
}
return CurrentFrequency;
}

void loop()
{
 int EndFreq = 0;
 EndFreq = FadeLED(RedPin, 0, 255, 1, 10, 300, 1);
 FadeLED(RedPin, 0, 255, -1,10 , EndFreq, 1);
}
```

Now upload the program to your Arduino. You should see the LED grow brighter as the tone generated by the buzzer increases in pitch. The LED will eventually reach its maximum and then start to fade down. At this point, the buzzer should start decreasing in pitch until the LED is fully off. Then the cycle will start again.

Hands-on Example: Arduino Random RGB LED Colors Using a Potentiometer

This hands-on example project shows you how to build a circuit that generates random colors that are displayed using an RGB LED. You can control the rate that the colors are generated with a potentiometer. The potentiometer allows you to gradually decrease or increase the rate that the colors change and are displayed using the LED.

Parts List

The parts you will need for this hands-on example project include

- 1 potentiometer
- 1 RGB LED
- 1 breadboard
- 8 wires (approximately)

Setting Up the Hardware

Set up the hardware for this project by:

- Connecting a wire from pin 11 on the Arduino to the leg of the LED that handles the blue color input.
- Connecting a wire from pin 10 on the Arduino to the leg of the LED that handles the green color input.
- Connecting a wire from pin 9 on the Arduino to the leg of the LED that handles the red color input.
- Connecting a wire from one of the GND pins on the Arduino to the leg of the LED that serves as the GND or common.
- Connecting a wire from one of the outermost pins on the potentiometer to the 5-V output on the Arduino.
- Connecting a wire from the other of the outermost pins on the potentiometer to the GND on the Arduino.
- Connecting a wire from the center pin of the potentiometer to pin A0 on the Arduino (Figure 3-20).

Setting Up the Software

This hands-on example project requires you to use the built-in random(MaxColor) function that returns a random number between 0 and 1 less than the MaxColor input parameter.

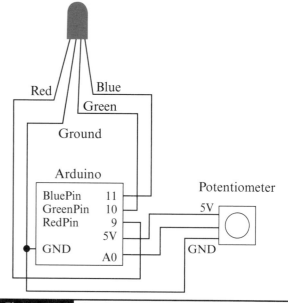

Figure 3-20 Random RGB color generator.

For example, calling random(256) returns a random value between 0 and 255. Another built-in Arduino function you will need is the analogRead(SensorPin) function, which maps an input voltage between 0 and 5 V to a number between 0 and 1,023. For example, the function analogRead(A0) reads the voltage at pin A0 on the Arduino and returns a number between 0 and 1,023.

The setup() function initializes the Serial Monitor for use in debugging Arduino programs and prints out a message indicating that the program has started. The GenerateRandomRGBColor(int MaxColor) function generates a random color by using the random() function to select a number between 0 and MaxColor for the red, green, and blue values of the LED. The SetRGBLEDColor() function is used to write the red, green, and blue values for the LED that were obtained by calling the GenerateRandomRGBColor() function to the actual RGB LED by calling the analogWrite() function. The PrintOutRGBColors() function prints out the current red, green, and blue values for the RGB LED to the Serial Monitor.

The main program function loop:

1. Reads the voltage value present at pin A0 from the potentiometer and converts the voltage to a number between 0 and 1,023.

2. Prints out this value to the Serial Monitor.

3. Generates a random RGB value between 0 and 255 by calling the GenerateRandomRGBColor() function with 256 as a parameter.

4. Sets the RGB value of the LED to the red, green, and blue values generated in step 3 by calling the SetRGBLEDColor() function.

5. Prints out the red, green, and blue values of the LED to the Serial Monitor by calling the PrintOutRGBColors() function.

6. Suspends execution of the program by calling the delay(CurrentSensorValue) function. This has the effect of displaying the current color on the LED for CurrentSensorValue milliseconds. See Listing 3-4.

Listing 3-4   Random Colors Using a Potentiometer

```
int SensorPin = A0;
int CurrentSensorValue = 0;

// Define LED Color Pin Outputs
// RGB value limits are 0-255
int BluePin = 11;
int GreenPin = 10;
int RedPin = 9;
int Red = 0;
int Green = 0;
int Blue = 0;

void setup()
{
 Serial.begin(9600);
 Serial.println("INITIALIZING Random RGB L.E.D. Color Generator .................");
}

void GenerateRandomRGBColor(int MaxColor)
{
 Red = random(MaxColor);
 Green = random(MaxColor);
 Blue = random(MaxColor);
}

void SetRGBLEDColor()
{
 analogWrite(RedPin, Red);
 analogWrite(GreenPin, Green);
```

(continued on next page)

Listing 3-4 Random Colors Using a Potentiometer (*continued*)

```
 analogWrite(BluePin, Blue);
}

void PrintOutRGBColors()
{
 Serial.print("Red: ");
 Serial.print(Red);
 Serial.print(", ");

 Serial.print("Green: ");
 Serial.print(Green);
 Serial.print(", ");

 Serial.print("Blue: ");
 Serial.println(Blue);
}

void loop()
{
 CurrentSensorValue = analogRead(SensorPin);
 Serial.print("POT value: ");
 Serial.print(CurrentSensorValue);
 Serial.print(", ");

 GenerateRandomRGBColor(256);
 SetRGBLEDColor();
 PrintOutRGBColors();

 delay(CurrentSensorValue);
}
```

Upload the program to the Arduino, and start the Serial Monitor. The Serial Monitor should display the current value of the potentiometer and the current red, green, and blue values of the LED. Turn the knob on the potentiometer clockwise as far as it will go. This should be one extreme where the LED blinks very rapidly and changes colors very fast or blinks very slowly and changes colors very slowly. Turn the knob from one extreme to the other, and note how fast the colors change. There should be a smooth transition from one extreme to the other. That is,

the colors should gradually start to change faster or slower as you turn the knob from one extreme to the other.

Hands-on Example: Arduino RGB Light Switch

In this hands-on example project, you will learn how to construct a circuit that controls an RGB LED like a light switch, but instead of turning a light on and off, it generates random colors every time you press and then release the button.

Parts List

The parts you will need for this hands-on example project include

- 1 push button
- 1 10,000-Ω resistor
- 1 RGB LED
- 1 breadboard
- 8 male-to-male wires (minimum)

Setting Up the Hardware

Setting up the hardware for this project consists of:

- Connecting a wire from pin 11 of the Arduino to the leg of the LED that handles the blue color input.

- Connecting a wire from pin 10 of the Arduino to the leg of the LED that handles the green color input.

- Connecting a wire from pin 9 of the Arduino to the leg of the LED that handles the red color input.

- Connecting a wire from one of the GND pins of the Arduino to the leg of the LED that serves as the GND or common.

- Connecting one of the pins on the push button to the 5-V source on the Arduino.

- Connecting the other pin on the push button to a node that contains pin 7 of the Arduino and one end of a 10,000-Ω, or 10-kΩ, resistor.

- Connecting the other end of the resistor to a GND pin on the Arduino (Figure 3-21).

Setting Up the Software

For this hands-on example project, you will need to read the digital input from a push button, and to do this, you will need to use the `digitalRead(pin)` function. This function

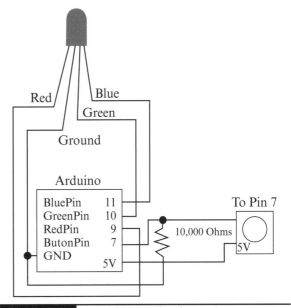

Figure 3-21 RGB LED switch schematic.

reads the voltage value at the pin specified. It returns a 1 if the voltage is HIGH, which means that it is 3 V or more, or a 0 if the voltage is LOW, or 2 V or less. The pin you use to read the voltage will also need to be declared as INPUT using the `pinMode` function, such as `pinMode(ButtonPin, INPUT)`. The first parameter is the pin number, and the second parameter is set to INPUT to indicate that this pin will be used to read the voltage value at that pin. Most of the code in this example project is reused from the preceding example project involving the potentiometer.

The main program loop:

1. Reads the voltage value at the pin connected to the push button.

2. If the result is low and the previous result was high, then the push button was just released. If this is true, then generate a random color, set this color as the color for the RGB LED, and then print out the red, green, and blue values of the color to the Serial Monitor. See Listing 3-5, and also note that the new code for this example is highlighted in bold.

Listing 3-5 The RGB Light Switch

```
// Define LED Color Pin Outputs
// RGB value limits are 0-255
int BluePin = 11;
int GreenPin = 10;
int RedPin = 9;
int ButtonPin= 7;

int Red = 0;
int Green = 0;
int Blue = 0;

int PreviousButtonPress = 0;

void setup()
{
 Serial.begin(9600);
 Serial.println("INITIALIZING RGB Light Switch Sketch...........................");
 pinMode(ButtonPin, INPUT);
}

void GenerateRandomRGBColor(int MaxColor)
{
 Red = random(MaxColor);
 Green = random(MaxColor);
 Blue = random(MaxColor);
}

void SetRGBLEDColor()
{
 analogWrite(RedPin, Red);
 analogWrite(GreenPin, Green);
 analogWrite(BluePin, Blue);
}

void PrintOutRGBColors()
{
 Serial.print("Red: ");
 Serial.print(Red);
 Serial.print(", ");

 Serial.print("Green: ");
 Serial.print(Green);
 Serial.print(", ");

 Serial.print("Blue: ");
 Serial.println(Blue);
}
```

```
void loop()
{
 int ButtonPressed = 0;
 ButtonPressed = digitalRead(ButtonPin);
 if ((PreviousButtonPress == 1) &&
     (ButtonPressed == 0))
 {
    // Button just released so process button
    GenerateRandomRGBColor(56);
    SetRGBLEDColor();
    PrintOutRGBColors();
 }
 PreviousButtonPress = ButtonPressed;
}
```

Now upload the program to your Arduino, and start the Serial Monitor. You should see the message initializing the program. Press and release the button, and a random color should be displayed on the LED. You should also see a message in the Serial Monitor that displays red, green, and blue values of the color. Each time you press and release the button, a new random color will be generated and displayed on the LED.

Hands-on Example: Raspberry Pi LED Blinker Counter

In this hands-on example project, you will learn how to set up a blinking LED on a Raspberry Pi that will blink on and off for a specified number of times before stopping. As each blink by the LED finishes, a message is printed to the Raspberry Pi's output terminal.

Parts List

The parts you will need for this hands-on example project include

- 1 LED
- 1 10,000-Ω resistor
- 1 breadboard

Setting Up the Hardware

You can set up the circuit by:

- Connecting a ground (GND) pin of the Raspberry Pi to the negative terminal of the LED.

- Connecting the positive terminal of the LED to one end of a 10,000-Ω resistor.

- Connecting the other end of the 10,000-Ω resistor to GPIO pin 14 on the Raspberry Pi (Figure 3-22).

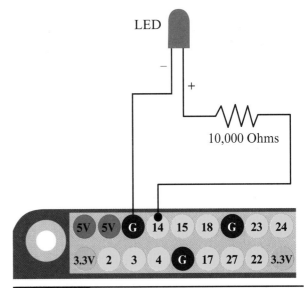

Figure 3-22 Blinking an LED on a Raspberry Pi.

Setting Up the Software

This example project uses the Rpi.GPIO library, which should be included by default with the Raspberry Pi 3 operating system. The code for this example project:

1. Imports the RPi.GPIO library as GPIO so that the module can be accessed with a reference to GPIO instead of the full name.

2. Imports the time module so that the sleep() function can be used in the program.

3. Sets the pin on the Raspberry Pi that will be connected to the LED to GPIO pin 14.

4. Sets the Counter variable, which keeps track of how many LED blinks have occurred, to 0.

5. Sets the Target variable, which holds the number the Counter variable needs to reach before the LED stops blinking.

6. Sets the RPi.GPIO library to recognize BroadCom or BCM pin numbering by calling the GPIO.setmode(GPIO.BCM) function with GPIO.BCM as the input parameter. BCM pin numbering for the GPIO pins is what this book uses to identify the pins you will need to use for the example projects for the Raspberry Pi.

7. Sets the LEDpin pin as an OUTPUT pin that will be used to provide a voltage to the attached LED by calling the GPIO.setup(LEDpin, GPIO.OUT) function. The first parameter is the pin number that will be set. The second parameter sets the pin mode to an OUTPUT pin.

8. While the value of the Counter variable is less than or equal to the value of the Target variable, the program:

   a. Turns on the LED by calling the GPIO.output(LEDpin, GPIO.HIGH) function, which sets a high voltage to the pin connected to the LED. The program

then suspends execution for half a second by calling the time.sleep(0.5) function. This has the effect of leaving the LED on for half a second.

   b. Turns off the LED by calling the GPIO.output(LEDpin, GPIO.LOW) function, which sets a 0 voltage value to the pin connected to the LED. The program then suspends execution for half a second by calling the time.sleep(0.5) function. This has the effect of leaving the LED off for half a second.

   c. Prints a message in the Raspberry Pi's terminal window indicating the number of blinks that have occurred, starting at 0, by calling the print("Number Blinks: ", Counter) function.

   d. Increases the Counter variable by 1.

9. After the program finishes blinking the light, the GPIO pins need to be reset and resources related to setting the pins need to be reallocated by calling the GPIO.cleanup() function. See Listing 3-6.

Listing 3-6 LED Blinker Counter

```
import RPi.GPIO as GPIO
import time

LEDpin = 14 #Set GPIO pin 14 for the LED
Counter = 0
Target = 5

GPIO.setmode(GPIO.BCM)
GPIO.setup(LEDpin, GPIO.OUT)

while (Counter <= Target):
 GPIO.output(LEDpin, GPIO.HIGH)
 time.sleep(0.5)
 GPIO.output(LEDpin, GPIO.LOW)
 time.sleep(0.5)
 print("Number Blinks: ", Counter)
 Counter = Counter + 1
GPIO.cleanup()
```

Start up a terminal window on your Raspberry Pi desktop, and change the directory to where you have saved the program. Run the program by typing in "python filename.py" where "filename.py" is the program shown in the preceding listing. You should see the LED light blink on and off, and you should also see a new message printed in the terminal window after each light blink cycle. When there have been six cycles of the light blinking on and off, then the light will stop blinking and the program will exit.

Hands-on Example: Raspberry Pi LED Fading

In this hands-on example project, you will learn how to control the brightness of a LED using a Python program on the Raspberry Pi 3.

Parts List

The parts you will need for this hands-on example project include

- 1 LED
- 1 10,000-Ω resistor
- 1 breadboard

The resistor is used to limit the brightness of the LED. If you want a brighter LED, then use a lower-value resistor or no resistor at all.

Setting Up the Hardware

You can set up the circuit by:

- Connecting a ground pin on the Raspberry Pi to the negative terminal of the LED.
- Connecting the positive terminal of the LED to one end of a 10,000-Ω resistor.
- Connecting the other end of the 10,000-Ω resistor to GPIO pin 14 on the Raspberry Pi (Figure 3-23).

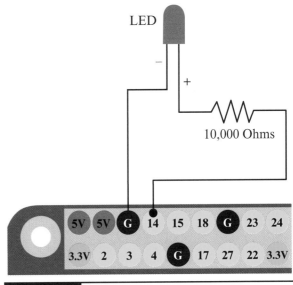

Figure 3-23 LED fading up and down.

Setting Up the Software

This example project uses PWM to simulate analog output using digital methods. In PWM, a square wave is generated that consists of digital 1s and 0s. The voltage output is always high or low, but when it is turned on and off very fast, this can simulate an analog output with a varying voltage output from 0 to 3.3 V. The *duty cycle* is the time that the voltage is high. By changing the duty cycle, we can change the LED's brightness. A duty cycle of 100 percent means that the LED is on all the time. A duty cycle of 50 percent means that the LED is on half the time, and a duty cycle of 0 percent means that the LED is always off.

The program in this example project:

1. Imports the RPi.GPIO and time modules so that we can use them in our program.
2. Connects GPIO pin 14 on the Raspberry Pi to the LED.
3. Sets the Freq variable that represents the frequency of the PWM for the LED to 50 hertz.
4. Sets the pin numbering to BCM numbering.

5. Sets the pin going to the LED to be an OUTPUT pin so that voltage can be output to the LED.

6. Sets up PWM using the Freq variable from step 3 (which is set at 50 hertz) on the pin connected to the LED.

7. Sets the PWM to 0 percent, or completely off.

8. Fades the LED up to full brightness and then down to completely off for six cycles. Prints a message to the terminal after each cycle is completed.

9. Stops the PWM.

10. Performs cleanup relating to the GPIO pin that was initialized. See Listing 3-7.

Listing 3-7 Fading a LED

```
# LED Fade Up and Fade Down
import RPi.GPIO as GPIO
import time

LEDpin = 14
Freq = 50

GPIO.setmode(GPIO.BCM)
GPIO.setup(LEDpin, GPIO.OUT)

p = GPIO.PWM(LEDpin, Freq)
p.start(0)

for counter in range(5):
    # Fade Up
    for dc in range(0, 100, 5):
        p.ChangeDutyCycle(dc)
        time.sleep(0.1)
    # Fade Down
    for dc in range(100, 0, -5):
        p.ChangeDutyCycle(dc)
        time.sleep(0.1)
    print("LED Fade Cycle: ", counter)
p.stop()
GPIO.cleanup()
```

After you type in or download this program from the publisher's website, run the program by going to the directory where the file is located and typing "python filename.py." This will run the program. You will see the LED fade up and then down, and after each cycle, you should see a message printed to the screen.

Hands-on Example: Raspberry Pi RGB LED Color Selector

In this hands-on example project, you will learn how to build a Raspberry Pi circuit that allows you to manually set the color of the RGB LED with the standard Raspberry Pi terminal using keyboard input.

Parts List

The parts you will need for this hands-on example project include

- 1 RGB LED
- 1 breadboard
- Male-to-female wires

Setting Up the Hardware

To build the circuit for this example project, you will need to:

- Connect the red terminal of the LED to GPIO pin 14 on the Raspberry Pi.
- Connect the ground terminal of the LED to a GND terminal on the Raspberry Pi.
- Connect the green terminal of the LED to GPIO pin 15 on the Raspberry Pi.
- Connect the blue terminal of the LED to GPIO pin 18 on the Raspberry Pi (Figure 3-24).

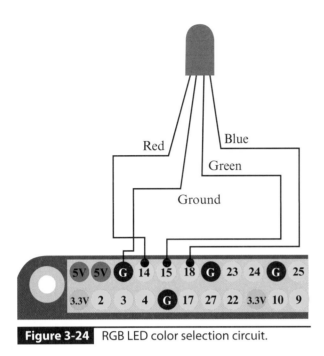

Figure 3-24 RGB LED color selection circuit.

Setting Up the Software

The program in this example project sets the RGB color of the RGB LED by:

1. Importing the RPi.GPIO and time modules into the program for use.

2. Defining pin 18 of the Raspberry Pi as the pin to connect to the blue terminal of the RGB LED.

3. Defining pin 15 of the Raspberry Pi as the pin to connect to the green terminal of the RGB LED.

4. Defining pin 14 of the Raspberry PI as the pin to connect to the red terminal of the RGB LED.

5. Initializing the variables that hold the user input values for the red, green, and blue colors to 0.

6. Setting the frequency used for the PWM for the RGB LED to 50 hertz.

7. Setting the pin numbering system to BCM.

8. Setting the pins on the Raspberry Pi that are connected to the red, green, and blue inputs on the RGB LED to output a voltage by calling the GPIO.setup() function.

9. Initializing the PWM functions on the Raspberry Pi pins connected to the red, green, and blue inputs on the LED by calling the GPIO.PWM() function for each pin.

10. Setting the PWM duty cycle on the Raspberry Pi pins connected to the red, green, and blue input pins on the LED to 0 percent, which means that they are turned completely off, by calling the start(0) function for each pin and setting the input parameter to 0.

11. Continuously asking the user to input an RGB value for the LED and then setting that color by:

    a. Asking the user to input a red value using the keyboard by calling the input("Enter Red Value (0-100): ") function and storing the returned user input. The input prompt message is sent as a parameter. If the returned value is less than 0, then the input while loop is exited.

    b. Repeating step 1 for green and blue, with the only difference being the input prompt message.

    c. Changing the duty cycle and thus brightness of the red, green, and blue components of the LED by calling the ChangeDutyCycle(pinnumber) function with the pin numbers of the pins connected to the red, green, and blue terminals on the RGB LED.

    d. Printing out the current color of the LED in red, green, and blue components to the terminal by using the print() function.

12. Discontinuing the PWM on the Raspberry Pi pins that provide the modulation by calling the `stop()` function.

13. Finally, freeing the resources related to the GPIO pins and resetting the pins that were used by calling the `GPIO.cleanup()` function. See Listing 3-8.

Listing 3-8   RGB LED Color Selector

```
# RGB LED Color Selector
import RPi.GPIO as GPIO
import time

# Define LED Color Pin Outputs
# RGB value limits are 0-100
BluePin = 18
GreenPin = 15
RedPin = 14
Red = 0
Green = 0
Blue = 0
Freq = 50

GPIO.setmode(GPIO.BCM)
GPIO.setup(RedPin, GPIO.OUT)
GPIO.setup(GreenPin, GPIO.OUT)
GPIO.setup(BluePin, GPIO.OUT)

PWMRed = GPIO.PWM(RedPin, Freq)
PWMGreen = GPIO.PWM(GreenPin, Freq)
PWMBlue = GPIO.PWM(BluePin, Freq)

PWMRed.start(0)
PWMGreen.start(0)
PWMBlue.start(0)

while 1:
   Red = input("Enter Red Value (0-100): ")
   if (Red < 0):
      break

   Green = input("Enter Green Value (0-100): ")
   if (Green < 0):
      break

   Blue = input("Enter Blue Value (0-100): ")
   if (Blue < 0):
      break

   PWMRed.ChangeDutyCycle(Red)
   PWMGreen.ChangeDutyCycle(Green)
```

```
    PWMBlue.ChangeDutyCycle(Blue)
    print("Current Color Value (R,G,B): ", Red, Green, Blue)

PWMRed.stop()
PWMGreen.stop()
PWMBlue.stop()
GPIO.cleanup()
```

After you have typed the program in or downloaded it from the publisher's website, start the program using Python. The program will ask you to enter red, green, and blue values for the LED. After you enter them, the values will be set on the LED. This loop continues until you enter a negative number for a color value, and then the program exits.

Hands-on Example: Raspberry Pi LED Fading Up and Down Using a Button

In this hands-on example, you will learn how to use a button to fade a LED up and down using pulse-width modulation.

Parts List

The parts you will need for this hands-on example project include

- 1 LED
- 2 10,000-Ω resistors
- 1 push button

Setting Up the Hardware

To set up the hardware for this example project, you will need to:

- Connect the negative terminal of the LED to GND on the Raspberry Pi.
- Connect the positive terminal of the LED to one end of a 10,000-Ω resistor.

- Connect the other end of the resistor to GPIO pin 14 on the Raspberry Pi.
- Connect one side of the push button to a 3.3-V pin on the Raspberry Pi.
- Connect the other end of the push button to a node that includes a 10,000-Ω resistor and GPIO pin 15 on the Raspberry Pi.
- Connect the other end of the resistor from the node to GND on the Raspberry Pi (Figure 3-25).

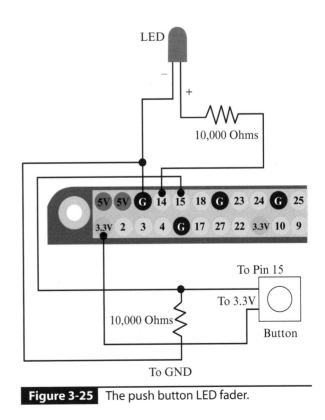

Figure 3-25 The push button LED fader.

Setting Up the Software

The program for this example project:

1. Imports the RPi.GPIO and time modules into the program so they can be used.

2. Assigns pin 14 of the Raspberry Pi to the LED.

3. Assigns pin 15 of the Raspberry Pi to the push button.

4. Resets the `Brightness` variable, which is used to keep track of the duty cycle of the LED, to 0.

5. Sets the `Delta` variable, which is the amount of change in the duty cycle per update, to 1.

6. Sets the frequency of the PWM for the LED to 50.

7. Sets the Raspberry Pi pin numbering system to BCM.

8. Sets the pin on the Raspberry Pi that is connected to the LED as an output pin to provide voltage to the LED.

9. Sets pin 15 on the Raspberry Pi, which is connected to the push button, as an input pin in order to read the voltage generated by a button press. A high reading indicates that the button is pressed, and a low reading indicates that the button is not pressed.

10. Initializes PWM on the Raspberry Pi pin connected to the LED and setting the duty cycle to 0, which means that the LED is turned completely off.

11. Reads the status of the push button by the Raspberry PI by calling the `GPIO.input(ButtonPin)` function, which returns the voltage level at the pin `ButtonPin`.

12. Updates the `Brightness` variable based on the value read from the push button.

13. Clamps the `Brightness` variable within the range of 0 (which is fully off) to 100 (which is fully on).

14. Changes the actual brightness of the LED by changing the duty cycle by calling the `LEDPWM.ChangeDutyCycle(Brightness)` function with the level of brightness as the input parameter.

15. Prints out the current brightness level of the LED to the terminal.

16. Suspends execution of the program for one-tenth of a second by calling the `time.sleep(0.1)` function with 0.1 as an input parameter. This has the effect of slowing down the changes in the brightness of the light. If this statement were left out, the changes in brightness would occur so rapidly that no fading up or down would be seen.

17. Returns program execution to step 11 until there is a keyboard exception such as the user pressing CTRL-C to halt execution of the program.

18. Stops PWM.

19. Resets the GPIO pins by calling the `cleanup()` function, and prints out a message to the terminal verifying that program execution has been halted. See Listing 3-9.

Listing 3-9 Push Button LED Fader

```
# Button LED Fade Up and Fade Down
# Press button to fade up LED and release button to
# fade down LED

import RPi.GPIO as GPIO
import time


LEDPin = 14
ButtonPin = 15
Brightness = 0
Delta = 1
Freq = 50

GPIO.setmode(GPIO.BCM)
GPIO.setup(LEDPin, GPIO.OUT)
GPIO.setup(ButtonPin, GPIO.IN)


LEDPWM = GPIO.PWM(LEDPin, Freq)
LEDPWM.start(0)

try:
 while 1:
    # Read in input from button
    ButtonPress = GPIO.input(ButtonPin)

    # Update LED brightness variable based on button push status
    if (ButtonPress):
       Brightness = Brightness + Delta
    else:
       Brightness = Brightness - Delta

    # Test LED brightness limits
    if (Brightness > 100):
       Brightness = 100
    if (Brightness < 0):
       Brightness = 0

    # Set actual LED Brightness
    LEDPWM.ChangeDutyCycle(Brightness)
    print("Current Brightness: ", Brightness)
    time.sleep(0.1)
except KeyboardInterrupt:
    pass
LEDPWM.stop()
GPIO.cleanup()
print("Exiting program ...")
```

Now you need to run the program. Either type the program in and save it or download it from the publisher's website. Run the program by typing "python filename.py" where the filename is the name of the program. The current brightness of the LED should be continuously displayed on the Raspberry Pi terminal. The brightness should start off as 0. Press the button to increase the brightness. You should see an increase in brightness of the LED. Release the button to decrease brightness. Press CTRL-C to exit the program.

Summary

This chapter has discussed basic electrical components and concepts relating to the Arduino and the Raspberry Pi. You learned basic concepts such as the circuit, series and parallel connections, resistors, and Ohm's law. Next, the color coding of resistors was presented, and you were given a chart that shows how to read a resistor's value. This was followed by a discussion of LEDs, RGB LEDs, piezo buzzers, potentiometers, push buttons, and breadboards. The rest of the chapter consisted of various hands on example projects involving the Arduino or Raspberry Pi using the components discussed earlier.

## CHAPTER 4

Touch Sensor Projects

THIS CHAPTER INTRODUCES YOU to the analog joystick and digital rotary encoder for the Arduino and Raspberry Pi. We start off by covering the analog joystick for the Arduino, and you learn how to connect and operate it. Next, we present a game using this joystick that is similar to the popular electronic game called "Simon," in which the user repeats a sequence of light patterns produced by a group of four LEDs. Then you learn about the rotary encoder controller. You learn how to connect and operate the encoder using an Arduino to control an RGB LED. This is followed by a discussion on how to connect and operate a rotary encoder using a Raspberry Pi. Finally, you use the Raspberry Pi to control a LED using the rotary encoder.

Figure 4-1 A joystick.

Analog Joystick

A *joystick* is a user input device that allows you to move a stick to the left or right and up or down. The joystick we will be using in this book is an analog joystick. That is, when a user moves the joystick, the values returned are within a range of values based on the position of the joystick (Figure 4-1).

Hands-on Example: Arduino Joystick Test

In this hands-on example, you will learn how to hook up an analog joystick to an Arduino and how to read the horizontal and vertical positions of the joystick.

Parts List

The parts you will need for this hands-on example project include

- 1 analog joystick
- 1 breadboard (optional)
- 4 wires (approximately)

Setting Up the Hardware

To set up the hardware for this hands-on example, you will need to:

- Connect the GND terminal of the joystick to the GND pin of the Arduino.
- Connect the 5-V terminal of the joystick to the 5-V pin of the Arduino.
- Connect the VRx pin on the joystick to the A0 or analog pin 0 of the Arduino.
- Connect the VRy pin on the joystick to the A1 or analog pin 1 of the Arduino (Figure 4-2).

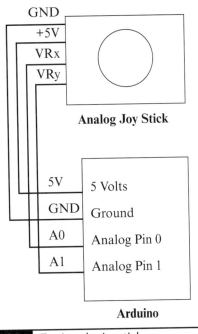

Figure 4-2   Testing the joystick.

Setting Up the Software

An analog joystick measures the horizontal and vertical positions of the joystick using two potentiometers, one on each axis. The values are between 0 and 1,023, assuming a 5-V input to the joystick. The leftmost position is 0, and the rightmost position is 1,023. The topmost position is 0, and the bottommost position is 1,023. The default center position is somewhere in between 0 and 1,023. In this example, the center position is read at the beginning of the program and is used to offset the raw input values from the Joystick. Thus the center position will be zero.

The Arduino program for this hands-on example:

1. Assigns the JoyPinX variable, which represents the VRx pin on the joystick, to analog pin 0 on the Arduino. This VRx pin represents the horizontal value of the joystick.

2. Assigns the JoyPinY variable, which represents the VRy pin on the joystick, to analog pin 1 on the Arduino. This VRy pin represents the vertical value of the joystick.

3. Locates the CenterX variable, which holds the adjusted x position of the joystick, as the center at 0.

4. Locates the CenterY variable, which holds the adjusted y position of the joystick, as the center at 0.

5. Initializes the X variable, which holds the raw value of the x position from the joystick.

6. Initializes the Y variable, which holds the raw value of the y position from the joystick.

7. Initializes the X1 variable, which holds the adjusted value of the x position from the joystick.

8. Initializes the Y1 variable, which holds the adjusted value of the *y* position from the joystick.

9. Defines the setup() function, which:

   a. Initializes the Serial Monitor to 9,600 baud and prints out a message indicating that the program has started.

   b. Determines the raw *x* and *y* values of the joystick center.

   c. Prints out the raw *x* and *y* values of the joystick center to the Serial Monitor.

10. Defines the main loop() function, which

    a. Reads the raw *x* and *y* joystick positions using the analogRead() function for the Arduino pins connected to the *x* and *y* joystick coordinate output pins.

    b. Calculates the adjusted *x* and *y* joystick coordinates, with (0,0) being the coordinates when the joystick is centered.

    c. Prints out the adjusted and raw *x* and *y* joystick coordinates to the Serial Monitor. See Listing 4-1.

Listing 4-1 Arduino Joystick Test

```
// Arduino Joystick Test
int JoyPinX = 0;
int JoyPinY = 1;

int CenterX = 0;
int CenterY = 0;

int X = 0;
int Y = 0;

int X1 = 0;
int Y1 = 0;

void setup()
{
  Serial.begin(9600);
  Serial.println("Arduino Joystick
    Test");

  CenterX = analogRead(JoyPinX);
  delay(100);
  CenterY = analogRead(JoyPinY);

  Serial.print("JoyStick Center: ");
  Serial.print(CenterX);
  Serial.print(" , ");
  Serial.println(CenterY);
}

void loop()
{
  // Reads X position of joystick
  X = analogRead(JoyPinX);

  // Wait before another analog read
  delay(100);

  // Reads Y position of joystick
  Y = analogRead(JoyPinY);

  // Calculate X,Y coords with Joystick
  // Center being 0,0
  X1 = X - CenterX;
  Y1 = Y - CenterY;

  // Prints out X,Y
  Serial.print(X1);
  Serial.print(" , ");
  Serial.print(Y1);

  Serial.print(" Raw (X,Y): ");
  Serial.print(X);
  Serial.print(" , ");
  Serial.println(Y);
}
```

Now you need to upload the program to the Arduino. After uploading, start the Serial Monitor. The first line should read

```
Arduino Joystick Test
```

This should be followed by the x and y coordinates of the joystick center in raw unadjusted values:

```
JoyStick Center: 508 , 522
```

When the joystick is centered, the output should look something like this. The first pair of numbers is the adjusted x and y coordinates of the joystick, and the last pair of numbers is the raw or unadjusted pair of x and y joystick coordinates (both sets of numbers are shown in bold print):

```
0 , 0 Raw (X,Y): 508 , 522
```

When the joystick is pushed all the way to the right, the x values (shown in bold print) should increase to their maximum:

```
515 , 0 Raw (X,Y): 1023 , 522
```

When the joystick is pushed all the way to the left, the x values (shown in bold print) should decrease to their minimum:

```
–508 , 0 Raw (X,Y): 0 , 522
```

When the joystick is pushed all the way up, the y values (shown in bold print) should decrease to their minimum:

```
0 , –522 Raw (X,Y): 508 , 0
```

When the joystick is pushed all the way down, the y values (shown in bold print) should increase to their maximum:

```
0 , 501 Raw (X,Y): 508 , 1023
```

If the x or y axis of the joystick is in between the minimum and maximum positions, then the x or y value should be in between the minimum and maximum values.

Hands-on Example: Arduino "Simon Says" Game

In this hands-on example project, you will learn how to build a game similar to the popular electronic game called "Simon" in which different colored lights flash in a certain order, and the player must reproduce this pattern by pressing buttons that correspond to the correct lights. However, in this project, you will be using a joystick as the input device that allows you to highlight a LED and then press a button to select this as an entry for the sequence that needs to be reproduced.

Parts List

To put together this hands-on example project, you will need

- 1 analog joystick
- 4 LEDs
- 4 10,000-Ω resistors (optional to reduce LED brightness)
- 1 push button
- 1 or more breadboards
- Wires to connect the components

Setting Up the Hardware

To set up the hardware for this hands-on example project, you will need to:

- Connect the GND terminal of the joystick to the GND pin of the Arduino.
- Connect the 5-V terminal of the joystick to the 5-V pin of the Arduino.
- Connect the VRx pin on the joystick to the A0 or analog pin 0 of the Arduino.
- Connect the VRy pin on the joystick to the A1 or analog pin 1 of the Arduino.

- Connect all the negative terminals on the LEDs to GND.

- Connect each of the positive terminals of the LEDS to a 10,000-Ω resistor.

- For one of the LEDs, connect the other end of the resistor to pin 11 of the Arduino.

- For one of the LEDs, connect the other end of the resistor to pin 10 of the Arduino.

- For one of the LEDs, connect the other end of the resistor to pin 9 of the Arduino.

- For one of the LEDs, connect the other end of the resistor to pin 8 of the Arduino.

- Connect one pin of the push button to the 5-V output node of the Arduino.

- Connect the other pin of the push button to a node that contains one end of a 10,000-Ω resistor. Then connect the other end of the resistor to GND.

- Connect pin 7 of the Arduino to the node that contains the 10,000-Ω resistor and the pin from the push button (Figure 4-3).

Setting Up the Software

The program that controls the hardware for this example consists of variables that hold the data needed for the program, functions that encapsulate key functions of the program, the setup() function that is called first when

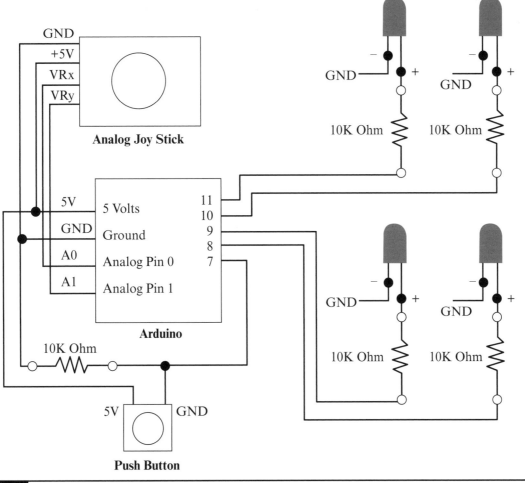

Figure 4-3 Arduino "Simon Says" game.

the Arduino is turned on, and the `loop()` function that is executed continuously by the Arduino after the `setup()` function has finished executing. This section discusses each of these key parts. First, we will look at the variables for the program. Next, we will discuss the `setup()` function . Then we will address the main `loop()` function and its supporting functions.

The `JoyPinX` and `JoyPinY` variables represent the pins on the joystick that output the *x* and *y* positions of the joystick. The pin that outputs the *x* position of the joystick is assigned to analog pin 0 of the Arduino. The pin that outputs the *y* position of the joystick is assigned to the analog pin 1 of the Arduino:

```
int JoyPinX = 0;
int JoyPinY = 1;
```

The `CenterX` and `CenterY` variables hold the raw *x* and *y* joystick positions of the joystick when the joystick is centered. Both of these variables are initialized to 0:

```
int CenterX = 0;
int CenterY = 0;
```

The `X` variable holds the adjusted value of the *x* joystick position, where a value of 0 is centered. The `Y` variable is not used in this example but is left for future expansion. Both variables are initialized to 0:

```
int X = 0;
int Y = 0;
```

The `ButtonPin` variable represents the push button that is connected to pin 7 of the Arduino:

```
int ButtonPin = 7;
```

The `PreviousButtonPress` variable is 1 if the button was pressed during the previous reading of the button and 0 if it was not pressed:

```
int PreviousButtonPress = 0;
```

The `LEDPin1` through `LEDPin4` variables represent the pin numbers on the Arduino to which each of these LEDs is connected:

```
int LEDPin1 = 8;
int LEDPin2 = 9;
int LEDPin3 = 10;
int LEDPin4 = 11;
```

The `NumberLEDS` variable holds the number of LEDs available to be used in the game and is initialized to 4:

```
const int NumberLEDS = 4;
```

The `MaxEntries` variable holds the maximum number of entries for a sequence and is set here to 5:

```
const int MaxEntries = 5;
```

The `NumberEntriesRequired` variable holds the number of entries that a player must enter for a sequence to be valid and is initialized to 4:

```
int NumberEntriesRequired = 4;
```

The `LEDNum2PinMap` array maps the LED number, such as 0 through 3, to the pin of the Arduino that is connected to that LED. The array holds a maximum of `NumberLEDS` elements:

```
int LEDNum2PinMap[NumberLEDS];
```

The `ComputerEntries` array holds the entries of the sequence generated by the Arduino that the player must try to match. The sequence is `MaxEntries` long:

```
int ComputerEntries[MaxEntries];
```

The `ComputerEntryGenerated` variable is true if a new Arduino-generated sequence has been created and false otherwise:

```
boolean ComputerEntryGenerated = false;
```

The `NumberUserEntries` variable holds the current number of selections the player has entered toward matching the Arduino-generated sequence:

```
int NumberUserEntries = 0;
```

The `UserEntries` array holds LED identification numbers such as 0 for LED number 0 that the user has selected so far in order to match the Arduino-generated sequence:

```
int UserEntries[MaxEntries];
```

The `UserSelection` variable holds the LED number that the player has chosen to enter into the player's matching sequence:

```
int UserSelection = 0;
```

The `JoystickThreshold` variable holds the minimum absolute value in the horizontal direction that is required to generate a joystick input event:

```
int JoystickThreshold = 150;
```

The `JoystickDelay` variable holds the minimum amount of time that must pass before two different joystick inputs are recognized by the program. This is done to slow down the speed at which the LED selection by the player is updated. The minimum time between recognizing two different joystick inputs is set to 500 milliseconds:

```
int JoystickDelay = 500;
```

The `JoystickPrevMovementTime` variable keeps track of the last time the joystick input was read:

```
unsigned long JoystickPrevMovementTime
= 0;
```

The `ButtonDelay` variable is the minimum time in milliseconds between reading input from the push button. This is needed so that electrical noise caused by pressing the button can settle before another reading can give accurate results. This process is called *debouncing*. The minimum delay is 350 milliseconds:

```
int ButtonDelay = 350;
```

The `ButtonLastToggledTime` variable keeps track of the last time the button changed state from being pressed to being released in milliseconds:

```
unsigned long ButtonLastToggledTime = 0;
```

The `Score` variable keeps track of the player's score:

```
int Score = 0;
```

The `EntryValue` is the value of each complete sequence. By default, a value of 10 is added to the player's score for each correct sequence of LEDs that he or she reproduces. A value of 10 is subtracted from the player's score for each incorrect sequence that the player enters:

```
int EntryValue = 10;
```

The `setup()` function initializes the program by:

1. Initializing the Arduino Serial Monitor and setting up the communication speed at 9,600 baud.

2. Printing out a message to the Serial Monitor indicating that the game has started.

3. Reading the *x* position of the joystick and setting this as the center *x* position.

4. Halting program execution for 100 milliseconds so that the next analog read is accurate.

5. Reading in the *y* position of the joystick and setting this as the center *y* position.

6. Printing out the center *x* and *y* positions to the Serial Monitor.

7. Setting the pin of the Arduino that is connected to the push button to be an INPUT pin that will read the voltage level.

8. Calling the `SetupLEDS()` function that initializes the LEDs that flash the sequence the player must reproduce. See Listing 4-2.

Listing 4-2 The setup() Function

```
void setup()
{
 Serial.begin(9600);
 Serial.println("Arduino Simon Says Game
  ...");

 CenterX = analogRead(JoyPinX);
 delay(100);
 CenterY = analogRead(JoyPinY);

 Serial.print("JoyStick Center: ");
 Serial.print(CenterX);
 Serial.print(" , ");
 Serial.println(CenterY);

 pinMode(ButtonPin, INPUT);
 SetupLEDS();
}
```

The SetupLEDS() function initializes the LEDs that form the sequence the player must replicate by:

1. Setting as OUTPUT pins the Arduino pins that are connected to the LEDs. These pins will provide the voltage to the LEDs to light them up when needed.

2. Calling the SetupLEDPinMapping() function to set up the system that maps LED identification numbers to actual Arduino pins. See Listing 4-3.

Listing 4-3 The SetupLEDS() Function

```
void SetupLEDS()
{
 pinMode(LEDPin1, OUTPUT);
 pinMode(LEDPin2, OUTPUT);
 pinMode(LEDPin3, OUTPUT);
 pinMode(LEDPin4, OUTPUT);

 SetupLEDPinMapping();
}
```

The SetupLEDPinMapping() function maps LED numbers to the actual Arduino pins to which the LEDs are connected using the LEDNum2PinMap array. See Listing 4-4.

Listing 4-4 The SetupLEDPinMapping() Function

```
void SetupLEDPinMapping()
{
 LEDNum2PinMap[0] = LEDPin1;
 LEDNum2PinMap[1] = LEDPin2;
 LEDNum2PinMap[2] = LEDPin3;
 LEDNum2PinMap[3] = LEDPin4;
}
```

The loop() function contains the main game logic and does the following:

1. If a new Arduino-generated sequence needs to be generated, then the function:

   a. Prints out a message to the Serial Monitor stating that you will need to press the button to start the game.

   b. Waits until the button has been pressed and then continues execution of the program.

   c. Generates a new sequence by calling the SelectComputerEntry() function.

   d. Prints out the new sequence to the Serial Monitor by calling the PrintComputerEntry() function.

   e. Displays the new sequence on the LEDs by calling the PlayComputerEntryOnLED() function.

   f. Prints a message to the Serial Monitor asking the player to hit the push button to start entering his or her matching sequence.

   g. Waits until the push button has been pressed before continuing program execution.

2. When the player begins his or her input, the program:

   a. Processes the player's joystick input by calling the `ProcessJoyStickMovement()` function.

   b. Displays the player's selection from step a on the correct LED by calling the `DisplayUserSelectionLED()` function.

3. If the player presses the button, then the program:

   a. Prints to the Serial Monitor the LED number that the player has selected.

   b. Adds the player's LED selection to the sequence that the player is generating.

   c. Flashes all the LEDs briefly by calling the `FlashAllLEDS()` function.

   d. Updates the number of entries that the player has entered for the sequence and print this number to the Serial Monitor.

4. If the number of entries the player has entered is equal to the required number of entries for the sequence to be completed, then the program:

   a. Turns off all the LEDs by calling the `TurnOffAllLEDS()` function.

   b. Tells the Arduino that another random sequence needs to be generated by setting the `ComputerEntryGenerated` variable to false.

   c. Resets the number of player entries for the sequence to 0.

   d. Tests to see if the player's sequence is the same as the Arduino-generated sequence by calling the `EvaluateUserEntry()` function.

5. If the player's sequence matches the Arduino's sequence, then the program prints out a message indicating the sequences matched, flashes the LEDs on and off with a long duration for the LEDs' on time and short off time, and updates the score. Otherwise, the program prints out a message indicating that the sequences did *not* match, flashes the LEDs on and off rapidly, and updates the score.

6. The program then prints the player's score to the Serial Monitor. See Listing 4-5.

Listing 4-5 The `loop()` Function

```
void loop()
{
  // Generate a new random entry if needed
  if(ComputerEntryGenerated == false)
  {
     Serial.println("Press Button to Start Game ...");
     while (ButtonPressed() == false)
     {
        // Do Nothing
     }

     SelectComputerEntry();
     PrintComputerEntry();
```

(continued on next page)

Listing 4-5 The loop() Function (*continued*)

```
      ComputerEntryGenerated = true;
      PlayComputerEntryOnLED();

      Serial.println("Press Button to Begin Your Entry ...");
      while (ButtonPressed() == false)
      {
         // Do Nothing
      }
   }


   // Process the joystick movement for LED selection
   ProcessJoyStickMovement();


   // Display User Selection on LEDS
   DisplayUserSelectionLED();


   // Test if entry button has been pressed
   if (ButtonPressed())
   {
      // Prints out User LED Selection
      Serial.print("User LED Selection: ");
      Serial.print(UserSelection);

      // Add user selection to list of user entries
      UserEntries[NumberUserEntries] = UserSelection;

      // Flash LEDS to indicate selection
      FlashAllLEDS(100, 100, 1);

      NumberUserEntries++;
      Serial.print(", Number User Entries: ");
      Serial.println(NumberUserEntries);

      if (NumberUserEntries >= NumberEntriesRequired)
      {
         TurnOffAllLEDS();
         ComputerEntryGenerated = false;
         NumberUserEntries = 0;

         boolean EntriesMatched = EvaluateUserEntry();
         if (EntriesMatched)
         {
            Serial.println("********** YES, Entries Matched **************");
            FlashAllLEDS(1000, 500, 6);
            Score += EntryValue;
         }
         else
         {
```

```
        Serial.println("----------- NO, DID NOT MATCH ---------------");
        FlashAllLEDS(100, 100, 6);
        Score -= EntryValue;
    }
    Serial.print("Score: ");
    Serial.println(Score);
  }
 }
}
```

The `SelectComputerEntry()` function generates a new sequence for the player to attempt to replicate by:

1. Generating a random number between 0 and 1 less than the number of LEDs availablefor each entry.

2. Adding this random number to the `ComputerEntries` array, which holds the computer-generated sequence that the player must reproduce. This number represents the LED number that the player must select to complete the sequence. See Listing 4-6.

Listing 4-6 The `SelectComputerEntry()` Function

```
void SelectComputerEntry()
{
 int randomnumber = 0;
 for (int i = 0; i <
   NumberEntriesRequired; i++)
 {
    randomnumber = random(NumberLEDS);
    ComputerEntries[i] = randomnumber;
 }
}
```

The `PrintComputerEntry()` function prints out each entry in the computer-generated sequence that is stored in the `ComputerEntries` array to the Serial Monitor. See Listing 4-7.

Listing 4-7 The `PrintComputerEntry()` Function

```
void PrintComputerEntry()
{
 Serial.print("Computer Entries:");

 for (int i = 0; i <
   NumberEntriesRequired; i++)
 {
    Serial.print(ComputerEntries[i]);
    Serial.print(" ");
 }
 Serial.println();
}
```

The `PlayComputerEntryOnLED()` function displays the Arduino-generated sequence that the player must reproduce by:

1. Obtaining the LED identification number from the `ComputerEntries` array.

2. Obtaining the Arduino pin number that is attached to this LED by using this identification number as an index into the `LEDNum2PinMap` array.

3. Lighting up the LED for 1,000 milliseconds.

4. Turning off the LED for 500 milliseconds. See Listing 4-8.

Listing 4-8 The `PlayComputerEntryOnLED()` Function

```
void PlayComputerEntryOnLED()
{
 int PinNumber = 0;
 int LEDNumber = 0;
 for (int i = 0; i <
  NumberEntriesRequired; i++)
 {
    LEDNumber = ComputerEntries[i];
    PinNumber =
      LEDNum2PinMap[LEDNumber];

    digitalWrite(PinNumber, HIGH);
    delay(1000);

    digitalWrite(PinNumber, LOW);
    delay(500);
 }
}
```

The `ButtonPressed()` function processes the player's button presses by:

1. Reading in the button status.

2. Then, if the previous button status was that the button was pressed and the current button status is that it has been released, the function calculates the time between this button release and the last one.

3. If the time between button releases is greater than the button delay time, then this is a legitimate button press. Otherwise, this button press may be from electrical noise generated from a previous button press and release and thus should be ignored. See Listing 4-9.

The `ProcessJoyStickMovement()` function processes the player's joystick input by:

1. Calculating the difference between the current time and the last time the joystick input was processed by the program.

Listing 4-9 The `ButtonPressed()` Function

```
boolean ButtonPressed()
{
 boolean result = false;
 int ButtonPressed = 0;

 ButtonPressed = digitalRead(ButtonPin);
 if ((PreviousButtonPress == 1) &&
     (ButtonPressed == 0))
 {
    unsigned long ButtonTimeBetweenToggle = millis() - ButtonLastToggledTime;
    if(ButtonTimeBetweenToggle > ButtonDelay)
    {
       // Button just released
       result = true;
       ButtonLastToggledTime = millis();
    }
 }
 PreviousButtonPress = ButtonPressed;

 return result;
}
```

2. Exiting if this difference is less than the joystick delay value.

3. Reading the adjusted *x* or horizontal position of the joystick by calling the `GetJoyStickX(false)` function with `false` as a parameter to indicate that we want the adjusted *x* position returned.

4. Updating the player's LED selection and making sure that the new LED number is within the valid range of LED numbers if the *x* joystick position is greater than the joystick threshold.

5. Updating the player's LED selection and making sure the new LED number is within the valid range of LED numbers if the *x* joystick position is less than the negative of the joystick threshold. See Listing 4-10.

Listing 4-10 The `ProcessJoyStickMovement()` Function

```
void ProcessJoyStickMovement()
{
  // Calc Time passed since last movement update
  unsigned int CurTime = millis();
  unsigned int Delay = CurTime - JoystickPrevMovementTime;
  if (Delay < JoystickDelay)
  {
     return;
  }

  // Reads X position of joystick and returns adjusted value
  X = GetJoyStickX(false);

  // if x axis joystick position is greater than threshold and joystick time delay
  // has passed then update User's LED selection
  if(X > JoystickThreshold)
  {
     UserSelection++;
     int MaxUserSelection = NumberLEDS - 1;
     if (UserSelection > MaxUserSelection)
     {
        UserSelection = MaxUserSelection;
     }
     JoystickPrevMovementTime = CurTime;
  }
  else
  if (X < -JoystickThreshold)
  {
     UserSelection--;
     if (UserSelection < 0)
     {
        UserSelection = 0;
     }
     JoystickPrevMovementTime = CurTime;
  }
}
```

The `GetJoyStickX()` function reads the *x* or horizontal position of the joystick. If the input parameter is true, the raw value that is within the range 0–1,023 assuming a 5-V input to the joystick is returned. If the input parameter is false, then an adjusted value of *x* is returned, where the *x* value of a centered joystick is 0. See Listing 4-11.

Listing 4-11 The `GetJoyStickX()` Function

```
int GetJoyStickX(boolean Raw)
{
 int result = 0;

 // Reads X position of joystick
 int x = analogRead(JoyPinX);

 // Calculate Adjusted x value
 int x1 = x - CenterX;

 // Return Raw or Adjusted Value
 if (Raw == true)
 {
    result = x;
 }
 else
 {
    result = x1;
 }
 return result;
}
```

The `DisplayUserSelectionLED()` function displays the player's current LED entry selection by:

1. Using the LED number that the player has selected as an index into the `LEDNum2PinMap` array to map the LED number to the actual pin of the Arduino to which the LED is connected and return the LED's pin number.

2. Turning off all the LEDs by calling the `TurnOffAllLEDS()` function.

3. Turning on the LED that the player has selected. See Listing 4-12.

Listing 4-12 The `DisplayUserSelectionLED()` Function

```
void DisplayUserSelectionLED()
{
 int PinNumber =
   LEDNum2PinMap[UserSelection];
 TurnOffAllLEDS();
 digitalWrite(PinNumber, HIGH);
}
```

The `TurnOffAllLEDS()` function turns off all the LEDs by:

1. Determining the pin number on the Arduino to which the LED is attached for each of the LEDs.

2. Turning off the LED by using the `digitalWrite()` function and setting the pin number obtained in the preceding step to a value of `LOW` for each of the LEDs. See Listing 4-13.

Listing 4-13 The `TurnOffAllLEDS()` Function

```
void TurnOffAllLEDS()
{
 int PinNumber = 0;
 for (int i = 0; i < NumberLEDS; i++)
 {
    PinNumber = LEDNum2PinMap[i];
    digitalWrite(PinNumber, LOW);
 }
}
```

The `FlashAllLEDS()` function flashes all the LEDs `NumLoops` times using the `HighTime` in milliseconds as the time to keep the LEDs on and the `LowTime` as the time to keep the LEDs off. See Listing 4-14.

Listing 4-14 The `FlashAllLEDS()` Function

```
void FlashAllLEDS(int HighTime, int
  LowTime, int NumLoops)
{
 int PinNumber = 0;

 for (int j = 0; j < NumLoops; j++)
 {
    for (int i = 0; i < NumberLEDS; i++)
    {
       PinNumber = LEDNum2PinMap[i];
       digitalWrite(PinNumber, HIGH);
    }
    delay(HighTime);

    for (int i = 0; i < NumberLEDS; i++)
    {
       PinNumber = LEDNum2PinMap[i];
       digitalWrite(PinNumber, LOW);
    }
    delay(LowTime);
 }
}
```

The `EvaluateUserEntry()` function compares each entry of the LED sequence entered by the player and the computer-generated LED sequence and returns true if the sequences match and false if the sequences do not match. See Listing 4-15.

Running the Program

Either type in the program or download it from the publisher's website. Upload the program to the Arduino, make sure that the joystick is centered, and run the Serial Monitor.

You should see a message indicating that the "Simon Says" game is now running along with the raw *x* and *y* values from the joystick that indicate the center values.

Listing 4-15 The `EvaluateUserEntry()` Function

```
boolean EvaluateUserEntry()
{
 boolean result = true;

 for (int i = 0; i <
   NumberEntriesRequired; i++)
 {
    if (UserEntries[i] !=
      ComputerEntries[i])
    {
       result = false;
       break;
    }
 }
 return result;
}
```

```
Arduino Simon Says Game ...
JoyStick Center: 508 , 522
```

You should also see a message indicating that you need to press the button to start the game. After you press the push button, the Arduino generates the sequence and displays it on the Serial Monitor and using the LEDs. You are prompted again to press the button to start your sequence entry. Press the button. You should see a LED light up. You can move the joystick left and right to move the LED selector over to the first LED in the sequence. Press the button to enter that LED as the first entry in the sequence. Repeat this for the rest of the LEDs in the sequence. After you enter the last LED in the sequence, the program validates the entry, and if it is correct, the score will be increased, all the LEDs should flash in long steady intervals, and a message saying that the entries match will be displayed on the Serial Monitor:

```
Press Button to Start Game ...
Computer Entries: 3 1 1 2
Press Button to Begin Your Entry ...
User LED Selection: 3, Number User
   Entries: 1
User LED Selection: 1, Number User
   Entries: 2
User LED Selection: 1, Number User
   Entries: 3
User LED Selection: 2, Number User
   Entries: 4
********** YES, Entries Matched
**************
Score: 10
```

After entering a few more correct entries, try to enter an incorrect sequence. For example, if the desired sequence is 0, 2, 3, 3 but you entered 0, 2, 3, 1, then the output displayed on the Serial Monitor would look something like this:

```
Press Button to Start Game ...
Computer Entries: 0 2 3 3
Press Button to Begin Your Entry ...
User LED Selection: 0, Number User
   Entries: 1
User LED Selection: 2, Number User
   Entries: 2
User LED Selection: 3, Number User
   Entries: 3
User LED Selection: 1, Number User
   Entries: 4
---------NO, DID NOT MATCH ------------
Score: 20
```

Rotary Encoder Controller (KY-040)

A rotary encoder controller is a device that can measure an infinite number of rotations in the clockwise and counterclockwise directions. The user either turns a shaft directly or turns a knob attached to the shaft to produce rotational motions that can be read by the Arduino or Raspberry Pi (Figure 4-4).

Figure 4-4 A rotary encoder.

Hands-on Example: Arduino Rotary Encoder Test

This hands-on example project shows the use of a rotary encoder with an Arduino. The encoder contains a shaft that you can turn clockwise or counterclockwise an infinite number of times. Some of these encoders also contain a shaft that can be used as a button. This device can be used for such things as a knob to allow the user to set a value in an Arduino program or in a game that uses a paddle that moves side to side across a thin-film-transistor (TFT) screen.

Parts List

To build this hands-on example project, you will need

- 1 rotary encoder
- 1 breadboard (optional, to stick the encoder into)
- 5 wires to attach the encoder to the Arduino

Setting Up the Hardware

To set up this example project, you need to:

- Connect the CLK (Output A) pin on the encoder to pin 7 of the Arduino.

- Connect the DT (Output B) pin on the encoder to pin 6 of the Arduino.

- Connect the SW pin on the encoder to pin 5 of the Arduino.

- Connect the + or VCC pin on the encoder to the 5- or 3.3-V pin of the Arduino.

- Connect the GND pin on the encoder to the GND pin of the Arduino (Figure 4-5).

Setting Up the Software

The program in this example reads in the output from the CLK (output A) and DT (output B) pins on the encoder, and based on the output signals, it determines whether the encoder is rotating and, if so, in what direction it is rotating. The rotating direction can be either clockwise (positive) or counterclockwise (negative). The program:

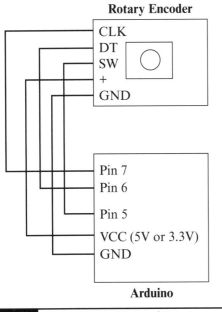

Rotary Encoder

CLK
DT
SW
+
GND

Pin 7
Pin 6
Pin 5
VCC (5V or 3.3V)
GND

Arduino

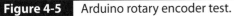

Figure 4-5   Arduino rotary encoder test.

1. Defines `EncoderPinA`, which represents the CLK pin from the encoder, as being connected to pin 7 on the Arduino.

2. Defines `EncoderPinB`, which represents the DT pin from the encoder, as being connected to pin 6 on the Arduino.

3. Defines `ButtonPin`, which represents the SW or Switch or Button pin from the encoder, as being connected to pin 5 on the Arduino.

4. Initializes the `EncoderPos` variable to 0, which is the encoder position where each movement of the encoder increments or decrements the position by 1.

5. Initializes the `PinAValueLast` variable to 0, which is the variable that holds the previous value read from output pin A from the last loop iteration.

6. Defines the `PinAValue` variable, which holds the current value read from output pin A on the encoder.

7. Defines the `PinBValue` variable, which holds the current value read from output pin B on the encoder.

8. Defines the `ButtonValue` variable, which holds the current value read from the SW pin on the encoder and is LOW or 0 if the button is being pressed.

9. Defines the `setup()` function, which:

   a. Sets as input the pins on the Arduino that are connected to the encoder's output A pin, output B pin, and SW or Button pin.

   b. Initializes the Serial Monitor and prints out a message indicating that the program has started.

   c. Reads the initial value from output pin A on the encoder by calling the `digitalRead(EncoderPinA)` function.

10. Defines the `loop()` function, which:

   a. Reads the values from the output A pin, the output B pin, and the SW pin on the encoder using the `digitalRead()` function.

   b. If the current value read from output pin A on the encoder is not equal to the last value read from output pin A, then:

      i. If the value of output pin A is the same as the value of output pin B, then the rotary encoder is being turned clockwise, so the function increments the encoder's position.

      ii. Otherwise, the rotary encoder is being turned counterclockwise, so the function decrements the encoder's position.

      iii. Prints out the encoder's position to the Serial Monitor.

   c. If the value for the button on the encoder is LOW or 0, then the button is being pressed.

   d. Saves the current value of output pin A in the `PinAValueLast` variable, which represents the last value read from output pin A. See Listing 4-16.

Listing 4-16 Arduino Rotary Encoder Test

```
// Arduino Rotary Encoder Test

int EncoderPinA = 7;
int EncoderPinB = 6;
int ButtonPin = 5;

int EncoderPos = 0;
int PinAValueLast = 0;
int PinAValue = 0;
int PinBValue = 0;
int ButtonValue = 0;

void setup()
```

```
{
  pinMode(EncoderPinA,INPUT);
  pinMode(EncoderPinB,INPUT);
  pinMode(ButtonPin, INPUT);

  Serial.begin (9600);
  Serial.println("Arduino Rotary Encoder
    Test... ");

  PinAValueLast = digitalRead(EncoderPinA);
}

void loop()
{
  PinAValue = digitalRead(EncoderPinA);
  PinBValue = digitalRead(EncoderPinB);
  ButtonValue = digitalRead(ButtonPin);

  if (PinAValue != PinAValueLast)
  {
    if (PinAValue == PinBValue)
    {
      // Clockwise
      EncoderPos++;
    }
    else
    {
      // Counter Clockwise
      EncoderPos--;
    }
    Serial.print("EncoderPos: ");
    Serial.println(EncoderPos);
  }

  if (ButtonValue == 0)
  {
    Serial.println(" Button Has Been
      Pressed ... ");
  }

  PinAValueLast = PinAValue;
}
```

Upload the program to your Arduino, and start the Serial Monitor. You should see the following message, which indicates that the `setup()` has been executed and the program has been initialized:

```
Arduino Rotary Encoder Test...
```

Next, turn the encoder shaft in the clockwise direction. You should see the encoder position increasing in the Serial Monitor as in:

```
EncoderPos: 1
EncoderPos: 2
EncoderPos: 3
EncoderPos: 4
```

Turn the encoder shaft in the counterclockwise direction. You should see the encoder's position decreasing in the Serial Monitor as in:

```
EncoderPos: 0
EncoderPos: -1
EncoderPos: -2
EncoderPos: -3
EncoderPos: -4
```

Finally, press down on the encoder shift until you hear a clicking sound. You should see messages that the button on the encoder has been pressed:

```
Button Has Been Pressed ...
Button Has Been Pressed ...
Button Has Been Pressed ...
```

Hands-on Example: Arduino Rotary Encoder Controlling an RGB LED

In this hands-on example project, you will use a rotary encoder to set, control, and display the red, green, blue, and composite values of an RGB LED. You will be able to adjust and set the red, green, and blue values of the RGB LED by rotating the rotary encoder clockwise and counterclockwise to raise or lower the intensity of the color and press down on the encoder to set the new value of the color. You will also be able display the composite RGB color on the LED.

Parts List

To perform this hands-on example project, you will need

- 1 rotary encoder
- 1 breadboard (optional, to stick the encoder into)
- Wires to attach the encoder and RGB LED to the Arduino
- 1 RGB LED

Setting Up the Hardware

To set up this hands-on example project, you need to:

- Connect the CLK (output A) pin on the encoder to pin 7 of the Arduino.
- Connect the DT (output B) pin on the encoder to pin 6 of the Arduino.
- Connect the SW pin on the encoder to pin 5 of the Arduino.
- Connect the + or VCC pin on the encoder to the 5- or 3.3-V pin of the Arduino.
- Connect the GND pin on the encoder to the GND pin of the Arduino.
- Connect the Red terminal on the LED to pin 9 of the Arduino.
- Connect the Ground terminal on the LED to a GND pin of the Arduino.
- Connect the Green terminal on the LED to pin 10 of the Arduino.
- Connect the Blue terminal on the LED to pin 11 of the Arduino (Figure 4-6).

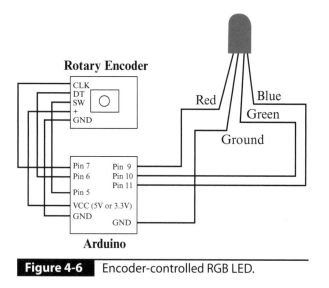

Figure 4-6   Encoder-controlled RGB LED.

Setting Up the Software

This hands-on example project builds on the last hands-on project, which tested out the rotary encoder. The new code is highlighted in bold type.

1. The `EncoderPosDelta` variable is the amount the color values are increased or decreased when the user turns the encoder shaft and is initialized to 1:

    ```
    int EncoderPosDelta = 1;
    ```

2. The `ButtonReleased` variable is true if the color selection button has just been released and needs to be processed. It is set to false after the button is processed by the program:

    ```
    boolean ButtonReleased = false;
    ```

3. The Red terminal on the RGB LED is assigned pin 9 of the Arduino:

    ```
    int RedPin = 9;
    ```

4. The Green terminal on the RGB LED is assigned to pin 10 of the Arduino:

    ```
    int GreenPin = 10;
    ```

5. The Blue terminal on the RGB LED is assigned to pin 11 of the Arduino:

    ```
    int BluePin = 11;
    ```

6. The `Red` variable holds the red component of the RGB LED:

    ```
    int Red = 10;
    ```

7. The `Green` variable holds the green component of the RGB LED:

    ```
    int Green = 10;
    ```

8. The `Blue` variable holds the blue component of the RGB LED:

    ```
    int Blue = 10;
    ```

9. The `Selections` enumeration holds the available options for the RGB LED. The `SetRed` selection sets the red component of the RGB LED. The `SetGreen` selection sets the green component of the RGB LED. The `SetBlue` selection sets the blue component of the RGB LED. The `ShowRGB` selection shows the composite of all the red, green, and blue values on the RGB LED:

    ```
    enum Selections
    {
    SetRed,
    SetGreen,
    SetBlue,
    ShowRGB
    };
    ```

10. The `Selection` variable holds the current active selection mode and is initialized to setting the red color component of the LED:

    ```
    Selections Selection = SetRed;
    ```

    See Listing 4-17. Note that this hands-on example is an extension of the rotary encoder test example and the code in bold print is new code added in.

Listing 4-17 Rotary Encoder RGB LED Controller

```
// Arduino Rotary Encoder and RGB LED
// Controller

// Encoder Pins
int EncoderPinA = 7;
int EncoderPinB = 6;
int ButtonPin = 5;

// Encoder Position
int EncoderPos = 0;
int EncoderPosDelta = 1;

int PinAValueLast = 0;
int PinAValue = 0;
int PinBValue = 0;

// Button
int ButtonValue = 0;
int ButtonPrevValue = 0;
boolean ButtonReleased = false;

// LED
int RedPin = 9;
int GreenPin = 10;
int BluePin = 11;

int Red = 10;
int Green = 10;
int Blue = 10;

enum Selections
{
  SetRed,
  SetGreen,
  SetBlue,
  ShowRGB
};
Selections Selection = SetRed;
```

The SetLEDColor(int R, int G, int B) function sets the red, green, and blue color components of the RGB LED using pulse-

width modulation (PWM) by calling the analogWrite() function. The first parameter is the Arduino pin on which to generate the PWM, and the second parameter is the value of the pulse in the range from 0 to 255. See Listing 4-18. Note that this function is completely new and not previously part of the rotary encoder test example. This is why the function is in all bold print.

Listing 4-18 The SetLEDColor() Function

```
void SetLEDColor(int R, int G, int B)
{
  // Input values 0-255
  analogWrite(RedPin, R);
  analogWrite(GreenPin, G);
  analogWrite(BluePin, B);
}
```

The setup() function initializes the program. This function is similar to the setup() function for the encoder test hands-on example covered previously. The changes to this function include the following:

1. The color of the RGB LED is set to the default red value, and all the other color components are set to 0. This indicates that the SetRed selection is active.

2. The encoder position is set, which indicates the user input to the value for the red component of the LED.

3. Some debug information is printed to the Serial Monitor. See Listing 4-19.

Listing 4-19 The `setup()` Function

```
void setup()
{
 pinMode(EncoderPinA,INPUT);
 pinMode(EncoderPinB,INPUT);
 pinMode(ButtonPin, INPUT);

 Serial.begin (9600);
 Serial.println("Arduino Rotary Encoder RGB LED Controller... ");

 PinAValueLast = digitalRead(EncoderPinA);
 ButtonPrevValue = digitalRead(ButtonPin);

 // Red Selection
 SetLEDColor(Red,0,0);
 EncoderPos = Red;
 PrintDebug();
}
```

The `ProcessRotaryEncoder()` function reads in both the rotational and push button input from the rotary encoder. This is the same function as you have seen before in the rotary encoder test example except that:

1. The rotary encoder position is adjusted by a variable called `EncoderPosDelta` instead of 1. This allows for better readability and allows the user to easily change the speed at which the color components are adjusted.

2. A new variable called `ButtonReleased` is set to true when the user has released the button on the encoder after having pressed it. See Listing 4-20.

Listing 4-20 The `ProcessRotaryEncoder()` Function

```
void ProcessRotaryEncoder()
{
 PinAValue = digitalRead(EncoderPinA);
 PinBValue = digitalRead(EncoderPinB);
 ButtonValue = digitalRead(ButtonPin);

 if (PinAValue != PinAValueLast)
 {
    if (PinAValue == PinBValue)
    {
       // Clockwise
       EncoderPos = EncoderPos + EncoderPosDelta;
    }
    else
    {
       // Counter Clockwise
```

```
      EncoderPos = EncoderPos - EncoderPosDelta;
   }
   Serial.print("EncoderPos: ");
   Serial.println(EncoderPos);
}


// For Button: LOW or 0 is pressed.
// HIGH or 1 is not pressed
if ((ButtonValue == 1) && (ButtonPrevValue == 0))
{
   Serial.println("Button Has Been Released ... ");
   ButtonReleased = true;
}


PinAValueLast = PinAValue;
ButtonPrevValue = ButtonValue;
}
```

The UpdateRGBLED() function sets the color of the RGB LED by:

1. Modifying the encoder position so that the values are within the range 0 to 255, which are the valid input values for a red, green, or blue color component.

2. Setting Red to the encoder position and setting the LED to this red color if the user is selecting the value of the red component.

3. Repeating step 2 for the green and blue color components.

4. Displaying the current colors held in the Red, Green, and Blue variables on the RGB LED by calling the SetLEDColor() function if the current selection is to show the composite RGB color on the LED.

5. Printing an error to the Serial Monitor if the current selection is none of the previous values. See Listing 4-21. Note that the entire function is in bold print because it is completely new and not previously part of the rotary encoder test example.

Listing 4-21 The UpdateRGBLED() Function

```
void UpdateRGBLED()
{
 // Clamp Encoder Values between 0-255 for RGB LED
 EncoderPos = constrain(EncoderPos, 0, 255);

 if(Selection == SetRed)
 {
 Red = EncoderPos;
 SetLEDColor(Red, 0, 0);
 }
 else
 if(Selection == SetGreen)
```

(continued on next page)

```
{
Green = EncoderPos;
SetLEDColor(0,Green,0);
}
else
if (Selection == SetBlue)
{
Blue = EncoderPos;
SetLEDColor(0,0,Blue);
}
else
if (Selection == ShowRGB)
{
SetLEDColor(Red, Green, Blue);
}
else
{
Serial.println("ERROR ... Selection not valid ...");
}
}
```

The `ProcessButtonRelease()` function processes the encoder selection button that enters the new red, green, or blue color for the LED or displays the composite RGB colors on the LED by:

1. Setting the new selection value to set the green color component and restoring the current value for the green component into the rotary encoder position variable if the current selection is setting the red component of the LED.

2. Setting the new selection value to set the blue color component and restoring the current value of the blue component into the rotary encoder position variable if the current selection is setting the green component of the LED.

3. Displaying the new selection value as the composite of the red, green, and blue components of the LED if the current selection is setting the blue component of the LED.

4. Setting the new selection value to set the red component of the LED and restoring the red component value back into the rotary encoder position variable if the current selection is to show the composite RGB LED color. See Listing 4-22. Note that the entire function is in bold because it is completely new and not part of the previous hands on example.

Listing 4-22 The `ProcessButtonRelease()` Function

```
void ProcessButtonRelease()
{
    // Move to next selection.
    if(Selection == SetRed)
    {
        Selection = SetGreen;
        EncoderPos = Green;
    }
    else
    if(Selection == SetGreen)
    {
        Selection = SetBlue;
        EncoderPos = Blue;
```

```
    }
    else
    if (Selection == SetBlue)
    {
        Selection = ShowRGB;
    }
    else
    if (Selection == ShowRGB)
    {
        Selection = SetRed;
        EncoderPos = Red;
    }
}
```

The `Selection2String()` function returns a string corresponding to the current selection mode, which is either setting the red, green, or blue components or showing the composite LED with all red, green, and blue components. See Listing 4-23. Note that the function is in all bold print because it is new and was not previously part of the rotary encoder test example.

Listing 4-23 The `Selection2String()` Function

```
String Selection2String(int s)
{
  String retvalue = "ERROR";

  switch(s)
  {
  case SetRed:
  retvalue = "Red";
  break;

  case SetGreen:
  retvalue = "Green";
  break;

  case SetBlue:
  retvalue = "Blue";
  break;

  case ShowRGB:
  retvalue = "RGB Composite";
  break;
  }
  return retvalue;
}
```

The `PrintDebug()` function prints the current selection mode and the current value of the LED's red, green, and blue values to the Serial Monitor. See Listing 4-24. Here is another function that is completely new to this hands-on example and thus in all bold print.

Listing 4-24 The `PrintDebug()` Function

```
void PrintDebug()
{
  Serial.print("Current Selection: ");
  Serial.print(Selection2String
    (Selection));

  Serial.print(" RGB: ");
  Serial.print(Red);
  Serial.print(" , ");
  Serial.print(Green);
  Serial.print(" , ");
  Serial.println(Blue);
}
```

The `loop()` function executes continuously and performs the main encoder controller logic by:

1. Reading in the angular rotation amount and the push button status of the rotary encoder device.

2. Updating the color on the RGB LED according to the current selection mode and current values of the red, green, and blue color components.

3. If the button has been pressed and then released, then the function:

   a. Processes the user's selection based on the selection mode.

   b. Resets the `ButtonReleased` variable to false so that another button press and release can be detected.

   c. Prints some informational debug statements to the Serial Monitor that indicate the current selection mode and the current values for the red, green, and blue components of the LED.

See Listing 4-25.

Listing 4-25 The `loop()` Function

```
void loop()
{
  // Process Rotary Encoder Input
  ProcessRotaryEncoder();

  // Update RGB LED
  UpdateRGBLED();

  // If button has been released then
  // process button
  if (ButtonReleased)
  {
  ProcessButtonRelease();
  ButtonReleased = false;
  PrintDebug();
  }
}
```

Running the Program

Upload the program to your Arduino. Start the Serial Monitor, and you should see the initialization message, the current selection mode, and the current red, green, and blue values for the LED:

```
Arduino Rotary Encoder RGB LED
  Controller...
Current Selection: Red RGB: 10 , 10 , 10
```

The default red, green, and blue values are 10 each, and the initial selection mode defaults to selecting the red component value. Try to rotate the shaft on the encoder, and you should see the encoder position increase or decrease depending on the direction you rotate the shaft:

```
EncoderPos: 9
EncoderPos: 8
```

This encoder position is used to set the red, green, and blue color components. Now press and release the encoder shaft. You should see

messages on the Serial Monitor similar to the following:

```
Button Has Been Released ...
Current Selection: Green RGB: 8 , 10 , 10
```

The selection mode should now be green for selecting the green component of the color using the rotary encoder. Twist the shaft to raise or lower the green component's value, and then press and release the shaft to move on to setting the blue component value:

```
Button Has Been Released ...
Current Selection: Blue RGB: 8 , 2 , 10
```

Try to raise and lower the value of the blue component by turning the shaft. Press and release the button again, and you will have selected the RGB composite mode that shows all the red, green, and blue colors combined on the LED:

```
Button Has Been Released ...
Current Selection: RGB Composite RGB:
  8 , 2 , 18
```

Hands-on Example: Raspberry Pi Rotary Encoder Test

In this hands-on example project, you will learn to operate a rotary encoder with a Raspberry Pi. This example project shows you how to measure the angular movement of the rotary encoder's shaft and introduces you to the push button feature, where the shaft of the encoder acts as a regular push button.

Parts List

The components needed for this example project include

- 1 rotary encoder

- 1 breadboard (optional)

- 5 wires to connect the rotary encoder to the Raspberry Pi

Setting Up the Hardware

To connect the hardware needed for this hands-on example, you will need to:

■ Connect the CLK (output A) pin on the rotary encoder to pin 14 of the Raspberry Pi.

■ Connect the DT (output B) pin on the rotary encoder to pin 15 of the Raspberry Pi.

■ Connect the SW (push button) pin on the rotary encoder to pin 18 of the Raspberry Pi.

■ Connect the Vcc (+) pin on the rotary encoder to the 3.3-V pin of the Raspberry Pi.

■ Connect the GND pin on the rotary encoder to one of the GND pins of the Raspberry Pi (Figure 4-7).

Setting Up the Software

The software for this hands-on example:

1. Imports the Rpi.GPIO module into the program.

2. Assigns pin 14 of the Raspberry Pi to output pin A or CLK on the rotary encoder.

3. Assigns pin 15 of the Raspberry Pi to output pin B or DT on the rotary encoder.

4. Assigns pin 18 of the Raspberry Pi to the button pin or SW on the rotary encoder.

5. Initializes some key variables. The encoder position is initialized to 0, and the variable that is used to determine the amount the encoder position is incremented or decremented is set to 1.

6. Sets the pin numbering scheme to BCM mode.

7. Declares the Raspberry Pi pins that are connected to output A, output B, and the push button on the rotary encoder as input pins so that the voltage values on those pins can be read.

8. Reads the voltage value from the output A pin on the encoder and stores it in the `PinAValueLast` variable.

9. Performs the following as long as the user does not execute a key board interrupt:

   a. Reads in the values from output A pin, output B pin, and the push button pin from the encoder.

   b. If the present value of output pin A is not the same as the last value of output pin A that was read in, then:

      i. Increases the encoder's position by `Delta` amount if the values from output pin A and output pin B are the same.

      ii. Otherwise, decrements the value of the encoder's position by `Delta` amount.

      iii. Prints the encoder's position to the terminal.

   c. If the button value is 0 or LOW voltage, then the button is being pressed, so a message is printed to the terminal.

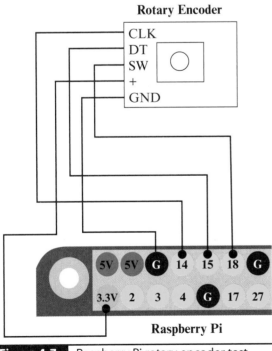

Rotary Encoder

CLK
DT
SW
+
GND

5V | 5V | G | 14 | 15 | 18 | G
3.3V | 2 | 3 | 4 | G | 17 | 27

Raspberry Pi

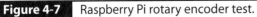

Figure 4-7 Raspberry Pi rotary encoder test.

d. The `PinAValueLast` is assigned the value that was just read in from the output pin A on the encoder.

e. Prints a message to the terminal indicating that the program is exiting.

10. Deallocates the GPIO resources by calling the `GPIO.cleanup()` function. See Listing 4-26.

Listing 4-26 Raspberry Pi Rotary Encoder Test

```
# Raspberry PI Rotary Encoder Test
import RPi.GPIO as GPIO

EncoderPinA = 14
EncoderPinB = 15
ButtonPin = 18

EncoderPos = 0
PinAValueLast = 0
PinAValue = 0
PinBValue = 0
ButtonValue = 0
Delta = 1
# Setup
GPIO.setmode(GPIO.BCM)
GPIO.setup(EncoderPinA, GPIO.IN)
GPIO.setup(EncoderPinB, GPIO.IN)
GPIO.setup(ButtonPin, GPIO.IN)

print("Raspberry PI Rotary Encoder
   Test... ")
PinAValueLast = GPIO.input(EncoderPinA)

try:
 while 1:
    PinAValue = GPIO.input(EncoderPinA)
    PinBValue = GPIO.input(EncoderPinB)
    ButtonValue = GPIO.input(ButtonPin)
    if (PinAValue != PinAValueLast):
       if (PinAValue == PinBValue):
          # Clockwise
          EncoderPos = EncoderPos +
             Delta
       else:
          # Counter Clockwise
          EncoderPos = EncoderPos -
             Delta
```

```
          print("EncoderPos: ", EncoderPos)
       if (ButtonValue == 0):
          print("Button Has Been Pressed
             ... ")
       PinAValueLast = PinAValue
except KeyboardInterrupt:
 pass
print("Exiting Rotary Encoder Test ...")
GPIO.cleanup()
```

Running the Program

Start the program by typing "python filename.py" in the directory in which you have stored the python program. Turn the encoder shaft clockwise, and you should see the encoder position increasing. Turn the encoder shaft counterclockwise, and you should see the encoder position decreasing. Now press the encoder's shaft down, and you should see a message saying that the encoder's button is being pressed. This example gives you the basics you will need to create your own Raspberry Pi programs using the encoder.

Hands-on Example: Raspberry Pi Rotary Encoder LED Blinker

This hands-on example project extends the preceding project using the Raspberry Pi and rotary encoder. Here we add a LED that has two modes. The default mode is that the LED light intensity can be adjusted by turning the encoder shaft clockwise or counterclockwise and then selecting that intensity by pressing down on the encoder shaft, which acts as a button. This intensity is now saved, and the second mode is activated, which fades the LED down from this selected intensity and then back up to this selected intensity. When the LED is fully faded up to the selected intensity, the mode is changed back to the default, which allows the user to adjust and then select another LED light intensity to fade down and back up again.

Parts List

For this hands-on example project, you will need

- 1 rotary encoder
- 1 LED
- 1 breadboard (optional for encoder)
- Enough wires to connect the LED and encoder to the Raspberry Pi

Setting Up the Hardware

To connect the hardware needed for this hands-on example, you will need to:

- Connect the CLK (output A) pin on the rotary encoder to pin 14 of the Raspberry Pi.
- Connect the DT (output B) pin on the rotary encoder to pin 15 of the Raspberry Pi.
- Connect the SW (push button) pin on the rotary encoder to pin 18 of the Raspberry Pi.
- Connect the Vcc (+) pin on the rotary encoder to the 3.3-V pin of the Raspberry Pi.
- Connect the GND pin on the rotary encoder to one of the GND pins of the Raspberry Pi.
- Connect the negative terminal of the LED to a GND pin of the Raspberry Pi.
- Connect the positive terminal of the LED to pin 17 of the Raspberry Pi (Figure 4-8).

Setting Up the Software

This example builds on the last hands-on example using the rotary encoder. The changes and additions to the source code are shown in bold print. The following changes and additions have been made to the source code from the preceding example:

1. A new variable that holds the LED intensity level is set to 10.

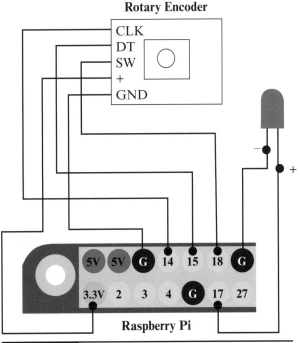

Figure 4-8 Rotary encoder LED controller.

2. Pin 17 on the Raspberry Pi is assigned to the positive terminal on the LED.

3. The frequency of the pulse-width modulation (PWM) on pin 17 is set to 50.

4. The selection mode is set to 0, which is the mode used to set the LED intensity by turning the encoder's shaft.

5. The button's previous value is set to 1, which means that the button was not pressed.

6. The variable that keeps track of user button presses is set to false.

7. The pin on the Raspberry Pi connected to the LED is set to be an output pin to provide voltage to the LED. This pin is also designated as outputting PWMs.

8. The default LED intensity is assigned to the encoder position, which can then be increased or decreased by the user.

9. A message is printed out to the terminal indicating that the program has been initialized.

10. A new function is created called `ProcessEncoderInput()`, which reads the encoder's position and the status of the button. Most of this code comes from the preceding hands-on example, but some additional code:

    a. Declares global variables "global" in the function, thus allowing them to be used inside the function.

    b. Limits the encoder position to a value within the range of 0 to 100.

    c. If the current button status is not pressed and the previous button status was pressed, then the program sets the variable that keeps track of the button press status to 1. This means that the button was just released and should be processed.

    d. Assigns the value of the current button status to the previous button status.

11. A new function that is called `def FadeDownUp()` is created that fades up and down the LED by:

    a. Fading the LED down to 0 from the LED intensity value in `Delta` increments.

    b. Fading the LED from 0 up to the LED intensity value in `Delta` increments.

    c. Printing a message to the terminal indicating the LED intensity value used.

    d. Setting `Mode` to 0, which is the mode that allows the user to change and then select the intensity of the LED he or she wants to have fade down and then back up.

12. The program does the following until the user generates a keyboard interrupt by pressing CTRL-C:

    a. Processes the rotary encoder input by calling the `ProcessEncoderInput()` function.

    b. If the encoder button is pressed, then:

       i. Sets the encoder position back to the LED intensity value if the current mode is fading the LED up and down.

       ii. Inverts the mode so that if the current mode is 0, then the new mode is 1, and vice versa.

       iii. Resets the `ButtonPressed` variable to 0 to indicate that the button press has been processed.

       iv. Assigns the encoder position to the LED intensity variable and changes the brightness of the LED using this new LED intensity if the mode is 0, which is the LED intensity selection mode.

       v. Fades the LED down and then back up by calling the `FadeDownUp()` function if the mode is not 0.

13. The program prints a message to the terminal indicating that the program is now exiting.

14. The program stops the PWM output to the pin on the Raspberry Pi that is connected to the LED by calling the `PWMLED.stop()` function. See Listing 4-27.

Listing 4-27 The Rotary Encoder LED Controller

```
# Raspberry PI Rotary Encoder LED Controller
import RPi.GPIO as GPIO
import time

EncoderPinA = 14
EncoderPinB = 15
ButtonPin = 18

EncoderPos = 0
PinAValueLast = 0
Delta = 1

# LED
LEDIntensity = 10
LEDPin = 17
Freq = 50
Mode = 0
ButtonPrevValue = 1
ButtonPressed = 0

# Encoder Setup
GPIO.setmode(GPIO.BCM)
GPIO.setup(EncoderPinA, GPIO.IN)
GPIO.setup(EncoderPinB, GPIO.IN)
GPIO.setup(ButtonPin, GPIO.IN)
PinAValueLast = GPIO.input(EncoderPinA)

# LED setup
GPIO.setup(LEDPin, GPIO.OUT)
PWMLED = GPIO.PWM(LEDPin, Freq)
PWMLED.start(0)
EncoderPos = LEDIntensity
print("Raspberry PI Rotary Encoder LED Controller and Blinker... ")

def ProcessEncoderInput():
 global PinAValueLast
 global ButtonPrevValue
 global EncoderPos
 global ButtonPressed

 PinAValue = GPIO.input(EncoderPinA)
 PinBValue = GPIO.input(EncoderPinB)
 ButtonValue = GPIO.input(ButtonPin)
 if (PinAValue != PinAValueLast):
    if (PinAValue == PinBValue):
        # Clockwise
```

(continued on next page)

Listing 4-27   The Rotary Encoder LED Controller (*continued*)

```
        EncoderPos = EncoderPos + Delta
    else:
        # Counter Clockwise
        EncoderPos = EncoderPos - Delta

    # Limit Range to 0-100
    if (EncoderPos > 100):
        EncoderPos = 100
    elif(EncoderPos < 0):
        EncoderPos = 0
    print("EncoderPos: ", EncoderPos)
    # 0 is pressed and 1 is not pressed
    if ((ButtonValue == 1) and (ButtonPrevValue == 0)):
        print("Button Has Been Released ... ")
        ButtonPressed = 1
    PinAValueLast = PinAValue
    ButtonPrevValue = ButtonValue

def FadeDownUp():
    global Mode
    # Fade Down
    for dc in range(LEDIntensity, 0, -Delta):
        PWMLED.ChangeDutyCycle(dc)
        time.sleep(0.05)
    # Fade Up
    for dc in range(0, LEDIntensity, Delta):
        PWMLED.ChangeDutyCycle(dc)
        time.sleep(0.05)
    print("LEDIntensity: ", LEDIntensity)
    Mode = 0

try:
    while 1:
        ProcessEncoderInput()
        if (ButtonPressed == 1):
            if (Mode == 1):
                EncoderPos = LEDIntensity
            Mode = not Mode
            ButtonPressed = 0

        if (Mode == 0):
            # Select LED Intensity Mode
            LEDIntensity = EncoderPos
            PWMLED.ChangeDutyCycle(LEDIntensity)
        else:
            # Fade Down and Up LED Mode
            FadeDownUp()
```

```
except KeyboardInterrupt:
 pass
print("Exiting Rotary Encoder LED Controller ...")
PWMLED.stop()
GPIO.cleanup()
```

Running the Program

Run the program from the terminal by executing "python filename.py," where filename.py is the name of the program file. The LED should light up, and a message indicating that the program has been initialized should be printed to the terminal. The program is now in mode 0. Turn the encoder shaft, and you should see the encoder position increase or decrease. Change the intensity of the LED light, and then press and release the rotary encoder shaft to select this intensity for the next mode. The next mode fades the LED down from the current intensity and then fades it back up to the same intensity. Next, the program mode should now switch back to mode 0.

Summary

This chapter has introduced the analog joystick and the rotary encoder for the Arduino and Raspberry Pi platforms. We started off by showing how to connect and operate an analog joystick with an Arduino. Then we built a game using the joystick for the Arduino. Then the rotary encoder for an Arduino was introduced, followed by an example of controlling an RGB LED with the encoder. Next, you learned how to use a rotary encoder with a Raspberry Pi. And finally, an example was presented where a rotary encoder was used with a Raspberry Pi to control a LED.

## CHAPTER 5

Environmental Sensor Projects I

THIS CHAPTER INTRODUCES YOU to many different types of environmental sensors for the Arduino and the Raspberry Pi. We start off by covering the reed switch. This is followed by hands-on example projects for the Arduino and Raspberry Pi using reed switches, such as an example that can be easily modified to be a door entry alarm. Next, TMP36 analog temperature sensor is introduced, and you learn how to measure temperature using this sensor with the Arduino. Flame sensors are next, and hands-on example projects using flame sensors are introduced as part of a fire alarm system for both the Arduino and Raspberry Pi. Next, you learn how to use an

infrared proximity or collision detector to sound an alarm when an object gets close to the sensor. The chapter closes with a demonstration of how to use a DHT11 temperature and humidity sensor with the Arduino.

Reed Switch Magnetic Field Sensor

A reed switch detects the presence of a magnet. The key feature of a reed switch is the cylindrical glass tube that detects the actual magnetic field from a magnet (Figure 5-1).

Figure 5-1   Reed switch.

Hands-on Example: Arduino Reed Switch Test

In this hands-on example, you will test a reed switch using the Arduino. A reed switch is used to detect the presence of a magnetic field.

Parts List

The parts you will need for this example project include

- 1 reed switch

- 1 breadboard (optional for reed switch)

- Wires for connecting the reed switch to the Arduino

Setting Up the Hardware

To set up the hardware for this project, you need to:

- Connect the VCC pin on the reed switch to the 3.3-V pin of the Arduino.

- Connect the GND pin on the reed switch to the GND pin of the Arduino.

- Connect the D0 pin on the reed switch to digital pin 7 of the Arduino (Figure 5-2).

Setting Up the Software

The software in this hands-on example project prints the magnetic field detection status of the reed switch by:

1. Assigning the D0 pin on the reed switch, which outputs the status of the magnetic field detection to digital pin 7 of the Arduino.

2. Initializing `RawValue`, which holds the value read from the reed switch to 0.

3. Defining the `setup()` function, which:

    a. Sets the mode of the pin that reads in data from the reed switch to INPUT.

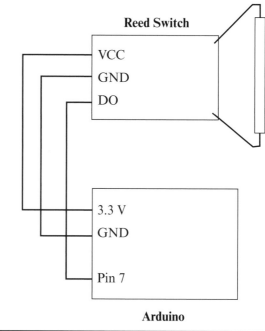

Figure 5-2   Arduino reed switch test.

    b. Initializes the Serial Monitor and prints out a message indicating that the program is starting.

4. Defining the `loop()` function, which:

    a. Reads the value from the reed switch.

    b. If this value is 0, then the reed switch has detected a magnetic field.

    c. If this value is 1, then the reed switch has not detected a magnetic field.

    d. Prints out the status of the magnetic field and the raw value read from the reed switch to the Serial Monitor. See Listing 5-1.

Running the Software

Upload the program to your Arduino, and start the Serial Monitor. You should see output indicating the status of the magnetic field detection and the associated raw value read from the reed switch. Take a magnet and bring it close to the reed switch. You should see the messages

Listing 5-1   The Arduino Reed Switch Test

```
// Reed Switch Test
int ReedSwitchPin = 7;
int RawValue = 0;

void setup()
{
 pinMode(ReedSwitchPin, INPUT);
 Serial.begin(9600);
 Serial.println("Reed Switch Test ....");
}

void loop()
{
 RawValue = digitalRead(ReedSwitchPin);
 if (RawValue == 0)
 {
    Serial.print("Reed Switch Detects Magnet ... ");
 }
 else
 {
    Serial.print("No Magnet Detected ...");
 }

 // 1 is off, 0 is activated by magnet
 Serial.print(" , RawValue: ");
 Serial.println(RawValue);
}
```

being printed to the Serial Monitor change to indicate that a magnetic field has been detected.

Hands-on Example: Arduino Door Buzzer Alarm

This hands-on example project adds to the preceding example project by adding a LED and a buzzer. The LED and buzzer activate when the reed switch detects a magnetic field and deactivate when the magnetic field is no longer there. This example project could be easily adapted for a simple door alarm that activates when one magnet on the door frame and another

magnet on the door itself go out of alignment, indicating that someone has opened the door.

Parts List

The parts required for this hands-on example project include

- 1 reed switch
- 1 LED
- 1 piezo buzzer
- 1 breadboard (optional)
- Wires to connect the components to the Arduino

Setting Up the Hardware

To set up the hardware for this example project, you need to

- Connect the VCC pin on the reed switch to the 3.3-V pin of the Arduino.

- Connect the GND pin on the reed switch to the GND pin of the Arduino.

- Connect the D0 pin on the reed switch to digital pin 7 of the Arduino.

- Connect the negative terminal on the LED to a GND pin of the Arduino.

- Connect the positive terminal on the LED to digital pin 8 of the Arduino.

- Connect the negative terminal on the buzzer to a GND pin of the Arduino.

- Connect the positive terminal on the buzzer to digital pin 9 of the Arduino (Figure 5-3).

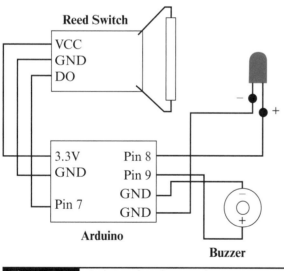

Figure 5-3 Arduino door buzzer alarm.

Setting Up the Software

The program in this example project builds on the preceding hands-on example. A LED and a buzzer have been added, and the code additions and changes are shown in bold type.

The new code additions include the following:

1. Pin 8 on the Arduino has been assigned to the positive terminal of the LED.

2. The on/off state of the LED has been initialized to off.

3. Pin 9 on the Arduino has been assigned to the positive terminal of the buzzer.

4. The frequency of the tone that the buzzer will produce is set to 300 hertz.

5. In the `setup()` function, the pin connected to the LED has been designated as an OUTPUT pin so that it can provide a voltage, and the initialization message has been updated.

6. In the main `loop()` function:

   a. If the reed switch detects a magnetic field, the LED state is set to on, and a tone is generated using the buzzer.

   b. If the Reed switch does not detect a magnetic field, the LED state is set to off, and any tone being generated by the buzzer is stopped.

   c. The LED is turned on or off based on the `LEDState` variable. See Listing 5-2.

Running the Program

Upload the program to your Arduino, and start the Serial Monitor. The output from the Arduino to the Serial Monitor should be the same as before, but when you bring the magnet near the reed switch, the buzzer and the LED should activate and stay activated as long as the magnet is detected. Pull the magnet away from the reed switch, and the LED and buzzer should turn off.

Listing 5-2 Arduino door buzzer alarm

```
// Reed Switch Door Buzzer
// Reed Switch
int ReedSwitchPin = 7;
int RawValue = 0;

// LED
int LEDPin = 8;
int LEDState = 0;

// Buzzer
int BuzzerPin = 9;
int BuzzerFreq = 300;

void setup()
{
 pinMode(ReedSwitchPin, INPUT);
 pinMode(LEDPin, OUTPUT);
 Serial.begin(9600);
 Serial.println("Reed Switch Door Buzzer ....");
}

void loop()
{
 RawValue = digitalRead(ReedSwitchPin);
 if (RawValue == 0)
 {
    Serial.print("Reed Switch Detects Magnet ... ");
    LEDState = 1;
    tone(BuzzerPin, BuzzerFreq);
 }
 else
 {
    Serial.print("No Magnet Detected ...");
    LEDState = 0;
    noTone(BuzzerPin);
 }

 // 1 is off, 0 is activated by magnet
 Serial.print(" , RawValue: ");
 Serial.println(RawValue);

 // Control LED
 digitalWrite(LEDPin,LEDState);
}
```

Hands-on Example: Raspberry Pi Reed Switch Test

In this hands-on example project, you will connect a reed switch to a Raspberry Pi.

Parts List

The components you will need for this hands-on example project include

- 1 reed switch
- 1 breadboard to hold the reed switch (optional)
- Wires to connect the reed switch to the Raspberry Pi

Setting Up the Hardware

To set up the hardware for this example project, you will need to:

- Connect the VCC pin on the reed switch to the 3.3-V pin of the Raspberry Pi.
- Connect the GND pin on the reed switch to the GND pin of the Raspberry Pi.
- Connect the D0 pin on the reed switch to pin 14 of the Raspberry Pi (Figure 5-4).

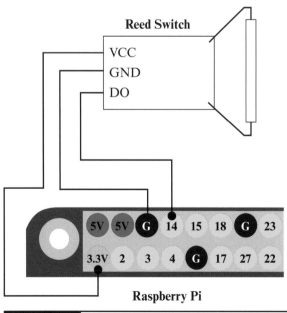

 Figure 5-4   Raspberry Pi reed switch test.

Setting Up the Software

The program for this hands-on example project demonstrates the basic use of a reed switch in detecting a magnetic field. The program:

1. Imports the RPi.GPIO library into the program to be accessed as GPIO.

2. Assigns pin 14 of the Raspberry Pi as the pin that will be connected to the output from the reed switch.

3. Initializes the variable that holds data read from the reed switch to 0.

4. Sets the pin numbering system of the Raspberry Pi to GPIO.BCM.

5. Sets the pin on the Raspberry Pi that is connected to the output pin on the reed switch to be an input pin that will read the value of the voltage at that pin.

6. Prints out a message to the terminal indicating that the program has been initialized.

7. Does the following until the user presses CTRL-C, which generates a keyboard interrupt:

    a. Reads the output value generated by the reed switch sensor.

    b. If this value is 0, then a magnetic field has been detected; otherwise, no magnetic field has been detected. Prints out a text message to the terminal indicating the result.

    c. Prints out the raw value read from the sensor to the terminal.

8. Prints out a text message to the terminal indicating that the program is exiting.

9. Deallocates resources related to the GPIO pins. See Listing 5-3.

Listing 5-3 Raspberry Pi Reed Switch Test

```
# Reed Switch Test
import RPi.GPIO as GPIO

ReedSwitchPin = 14
RawValue = 0

# Setup
GPIO.setmode(GPIO.BCM)
GPIO.setup(ReedSwitchPin, GPIO.IN)
print("Reed Switch Test ....")

try:
  while 1:
    RawValue = GPIO.input(ReedSwitchPin)
    if (RawValue == 0):
        print("Reed Switch Detects Magnet
            ... ")
    else:
        print("No Magnet Detected ...")

    # 1 is off, 0 is activated by magnet
    print(" , RawValue: ", RawValue)

except KeyboardInterrupt:
  pass
print("Exiting Reed Switch Test ...")
GPIO.cleanup()
```

Running the Software

Start the program on the Raspberry Pi. The current status of magnetic field detection and the corresponding raw value read from the reed switch should be displayed. Bring a magnet close to the reed switch. The readings on the terminal should change from indicating no magnetic field to indicating that a magnetic field is present. Pull the magnet away from the reed switch, and the original reading that no magnetic field is detected should return.

Hands-on Example: Raspberry Pi Reed Switch Door Alarm

In this hands-on example project, you will build on the preceding hands-on example project and learn how a basic door alarm that includes sound effects can be constructed.

Parts List

The components you will need for this hands-on example project include

- 1 reed switch
- 1 breadboard to hold the reed switch (optional)
- Wires to connect the reed switch to the Raspberry Pi
- Ability to output sound from the Raspberry Pi to a speaker

Setting Up the Hardware

To set up the hardware for this example project, you will need to:

- Connect the VCC pin on the reed switch to the 3.3-V pin of the Raspberry Pi.
- Connect the GND pin on the reed switch to the GND pin of the Raspberry Pi.
- Connect the D0 pin on the reed switch to pin 14 of the Raspberry Pi (Figure 5-5).

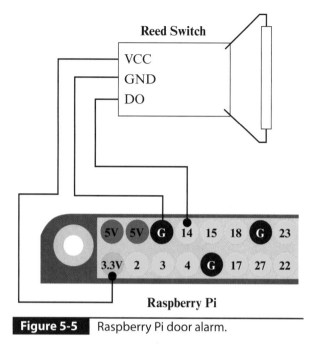

Figure 5-5   Raspberry Pi door alarm.

Setting Up the Software

This hands-on example project builds on the preceding example project. The key code additions and modifications to the program in this section are highlighted in bold type. The new feature that was added for this example is that of playing a sound effect whenever a magnetic field is detected. This involves the addition of code that:

1. Imports the pygame library into the program so that it can be used.

2. Initializes the pygame library.

3. Initializes the sound mixer.

4. Specifies a sound file and creates an object from this sound file called `Alarm`. The specific sound file we used is called `match1.wav` and is from one of the demo programs that came with the standard Raspberry Pi installation.

5. If a magnetic field is detected, then the program plays the sound effect created previously by calling the `Alarm.play()` function. See Listing 5-4.

Listing 5-4   Raspberry Pi Door Alarm

```
# Reed Switch Door Alarm
import RPi.GPIO as GPIO
import pygame
from time import sleep

ReedSwitchPin = 14
RawValue = 0

# Setup
GPIO.setmode(GPIO.BCM)
GPIO.setup(ReedSwitchPin, GPIO.IN)

#SFX
pygame.init()
pygame.mixer.init()
Alarm = pygame.mixer.Sound("match1.wav")
print("Reed Switch Door Alarm ....")

try:
 while 1:
    RawValue = GPIO.input(ReedSwitchPin)
    if (RawValue == 0):
      print("Reed Switch Detects Magnet
        ... ")
      Alarm.play()
    else:
      print("No Magnet Detected ...")

    # 1 is off, 0 is activated by magnet
    print(" , RawValue: ", RawValue)
except KeyboardInterrupt:
 pass
print("Exiting Reed Switch Test ...")
GPIO.cleanup()
```

Running the Program

Before running the program, you will need to determine what sound file you want for the program. The file I used was a WAV sound file I found in the pigames directory that was part of the default Raspberry Pi installation. You will need to copy whatever sound file you want to the same directory from which you are running the program. In addition, you will need to change the name of the sound file in the program to match the name of the sound file you just copied.

Start the program on the Raspberry Pi using Python. The current status of magnetic field detection and the corresponding raw value read from the reed switch should be displayed on the terminal. Bring a magnet close to the reed switch. The readings on the terminal should change from indicating no magnetic field to indicating that a magnetic field is present. In addition, a sound effect should start playing through your monitor's speaker. Pull the magnet away from the reed switch, and the readings when no magnetic field was detected should return. The sound effect that was previously playing should stop now.

TMP36 Temperature Sensor

The TMP36 temperature sensor provides an output voltage that is linearly proportional to the Celsius (Centigrade) temperature (Figure 5-6).

There are three pins on the TMP36. The +Vs pin should be connected to the 3.3-V pin of the Arduino. The center Vout pin outputs voltage that represents the measured temperature in

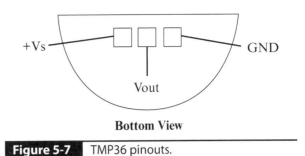

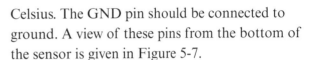

Figure 5-7   TMP36 pinouts.

Celsius. The GND pin should be connected to ground. A view of these pins from the bottom of the sensor is given in Figure 5-7.

The relationship between output voltage and temperature for the TMP36 is linear. This means that it can be modeled by the equation of a line:

$$y = mx + b$$

where y is the output voltage from the TMP36 sensor, x is the temperature in Celsius, m is the slope of the line, and b is the y-intercept value. From the data sheet for the TMP36 in Figure 5-6, which shows a graph of the voltage output versus temperature, you can see that the y-intercept value is 0.5. The slope of a line is the change between the y values divided by the change in x values between two points on the line or:

$$m = (y_2 - y_1)/(x_2 - x_1)$$

Two points on the line from Figure 5-6 in the data sheet for the TMP36 sensor are

$$pt1(0, 0.5) \quad \text{and} \quad pt2(50, 1)$$

The slope is then calculated as:

$$m = (1 - 0.5)/(50 - 0)$$
$$= 0.5/50 = 0.01 = 1/100$$

The equation of the line where y is the output voltage and x is temperature is:

$$y = x/100 + 0.5$$

To find x, which is temperature based on y, which is the output voltage, we need to solve the equation for the x variable as:

$$y - 0.5 = x/100$$

$$(y - 0.5) \times 100 = x = \text{temperature in Celsius}$$

Figure 5-6   TMP36 temperature sensor.

Thus we have a final equation that calculates the temperature in Celsius based on the output voltage from the TMP36 sensor as:

$$\text{Temperature} = (\text{Vout} - 0.5) \times 100$$

Hands-on Example: Arduino TMP36 Temperature Sensor Test

In this hands-on example project, you will learn how to connect and use a TMP36 temperature sensor to measure the temperature in Celsius and Fahrenheit using an Arduino.

Parts List

The components needed for this hands-on example project include

- 1 TMP36 temperature sensor
- 1 breadboard (optional for TMP36 sensor)
- Wires to connect the TMP36 sensor to the Arduino

Setting Up the Hardware

To set up the hardware for this example project, you will need to:

- Connect the +Vs pin on the TMP36 sensor to the 3.3-V output pin of the Arduino.
- Connect the GND pin on the TMP36 sensor to a GND pin of the Arduino.
- Connect the Vout pin on the TMP36 sensor to analog pin 0 of the Arduino (Figure 5-8).

Setting Up the Software

The program for this hands-on example project reads the temperature from the TMP36 sensor and displays the result in both Celsius and Fahrenheit by:

1. Assigning the analog pin 0 of the Arduino to the Vout pin on the TMP36.

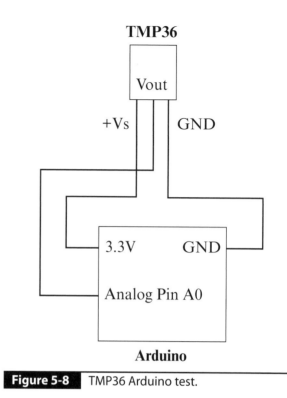

Figure 5-8   TMP36 Arduino test.

2. Initializing the variable that reads the raw value from the Vout pin to 0.

3. Initializing the variable that holds the voltage value read from the Vout pin to 0.

4. Defining the `UnitsToVolts` variable so that it converts units read by the Arduino from the Vout pin to voltage. The conversion rate is 5 V = 1,024 units.

5. Defining the `setup()` function as:

   a. Initializing the Serial Monitor with a speed of 9,600 baud.

   b. Printing out a message to the Serial Monitor indicating that the program has started.

6. Defining the `GetTempCelsius(float Voltage)` function and returning the temperature in Celsius based on the input parameter `Voltage`. The equation used to calculate the temperature is linear and was derived in the preceding section.

7. Defining the `GetTempFarenheit(float Voltage)` function and returning the

temperature in Fahrenheit based on the input parameter voltage. The `GetTempCelsius()` function is first called to get the temperature in Celsius, followed by conversion of the temperature from Celsius to Fahrenheit.

8. Defining the `loop()` function, which:

   a. Reads the raw value from the Vout pin of the TMP36.

   b. Converts the raw value from step 1 into a voltage value.

   c. Prints the temperature in Celsius to the Serial Monitor using the `GetTempCelsius(VoltageValue)` function, with the input parameter being the voltage value from step 2.

   d. Prints the temperature in Fahrenheit to the Serial Monitor using the `GetTempFarenheit(VoltageValue)` function, with the input parameter being the voltage value from step 2. See Listing 5-5.

Listing 5-5 Arduino TMP36 Test

```
// TMP 36 component test
int TMP36Pin = A0;
int RawValue = 0;
float VoltageValue = 0;

// Conversion: 5 volts/1024 units
const float UnitsToVolts = 5.0/1024.0;

void setup()
{
 Serial.begin(9600);
 Serial.println("Arduino TMP36 Test
   ...");
}

float GetTempCelsius(float Voltage)
{
 float temp = 0;
 temp = (Voltage-0.5)*100;
 return temp;
```

```
}

float GetTempFarenheit(float Voltage)
{
 float temp = 0;
 float CTemp = GetTempCelsius(Voltage);
 temp = CTemp*(9.0/5.0)+32.0;
 return temp;
}

void loop()
{
 RawValue = analogRead(TMP36Pin);
 VoltageValue = RawValue * UnitsToVolts;

 Serial.print("Celsius Temp: ");
 Serial.print(GetTempCelsius
   (VoltageValue));
 Serial.print(" , Farenheit Temp: ");
 Serial.println(GetTempFarenheit
   (VoltageValue));
}
```

Running the Software

Upload the program to your Arduino, and start the Serial Monitor. The temperature in Celsius and Fahrenheit should be displayed. You can try to slow down the temperature readings by adding a delay statement such as `delay(1000)` at the end of the `loop()` function. This delays the printing of the next temperature reading by 1,000 milliseconds, or 1 second. You can adjust the delay value to suit your own personal taste.

Flame Sensor

A flame sensor detects the presence of infrared heat that is associated with the heat from a fire or flame. A sensor like this could be used in an Arduino- or Raspberry Pi–based fire alarm (Figure 5-9).

Figure 5-9   A flame sensor.

Hands-on Example: Arduino Flame Sensor Test

In this hands-on example project, you will learn how to connect and use a flame sensor with your Arduino.

Parts List

The components you will need for this hands-on example project include

- 1 flame sensor
- 1 breadboard to hold the flame sensor
- Wires to connect the flame sensor to the Arduino

Setting Up the Hardware

To set up the hardware for this example project, you will need to:

- Connect the GND pin on the flame sensor to a GND pin of the Arduino.
- Connect the VCC pin on the flame sensor to the 3.3-V pin of the Arduino.
- Connect the DOUT pin on the flame sensor to digital pin 8 of the Arduino (Figure 5-10).

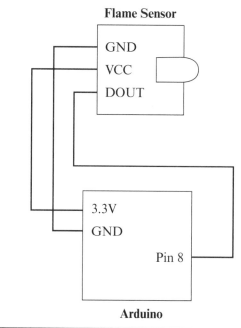

Figure 5-10   Flame sensor test.

Setting Up the Software

The program for this hands-on example project reads the flame sensor output and prints out the flame detection status to the Serial Monitor by:

1. Assigning digital pin 8 of the Arduino to the output pin on the flame sensor.

2. Initializing the variable that reads the raw value from the flame sensor to 0.

3. Defining the `setup()` function as:

   a. Setting the pin mode of the Arduino pin connected to the flame sensor's output pin to be an input pin in order to read the data from the sensor.

   b. Initializing the Serial Monitor and setting the transmission speed to 9,600 baud.

   c. Printing out a message to the Serial Monitor indicating that the program has started.

4. Defining the `loop()` function as:

   a. Reading in the output value from the flame sensor.

   b. If the value is equal to 0, then a flame has been detected, and the function prints out this information to the Serial Monitor.

   c. If the value is not equal to 0, then a flame has *not* been detected, and the function prints out this information to the Serial Monitor.

   d. Printing out the raw value read from the flame sensor to the Serial Monitor. See Listing 5-6.

Running the Program

Upload the program to your Arduino, and start the Serial Monitor. On the Serial Monitor you should see the flame detection status. Light a match or use a lighter and hold the flame in front of the black LED. You should see the flame detection status change from no flame

Listing 5-6 Flame Sensor Test

```
// Flame Sensor Test
int FlameOutPin = 8;
int RawValue = 0;

void setup()
{
 pinMode(FlameOutPin, INPUT);
 Serial.begin(9600);
 Serial.println("Flame Sensor Test ...");
}

void loop()
{
 RawValue = digitalRead(FlameOutPin);
 if (RawValue == 0)
 {
    Serial.print("Flame Detected !!!!!");
 }
 else
 {
    Serial.print("No Flame Detected ....");
 }
 Serial.print(" , RawValue: ");
 Serial.println(RawValue);
}
```

detected to flame detected. I was also able to activate the flame sensor using a flashlight.

Hands-on Example: Arduino Fire Alarm

In this hands-on example project, you will learn how to make a fire alarm using a flame sensor. This example builds on the preceding example project and adds in a buzzer that activates when the alarm has been tripped and a button that is used to reset the alarm.

Parts List

For this hands-on example project, you will need

- 1 flame sensor
- 1 piezo buzzer
- 1 push button
- 1 10-kΩ resistor
- 1 breadboard
- Wires to connect the components together

Setting Up the Hardware

To set up the hardware for this hands-on example project, you will need to:

- Connect the GND pin on the flame sensor to a GND pin of the Arduino.
- Connect the VCC pin on the flame sensor to the 3.3-V pin of the Arduino.
- Connect the DOUT pin on the flame sensor to digital pin 8 of the Arduino.
- Connect the GND terminal on the buzzer to a GND pin of the Arduino.
- Connect the positive terminal on the buzzer to digital pin 9 of the Arduino.
- Connect one terminal of the push button to the 3.3-V pin of the Arduino.

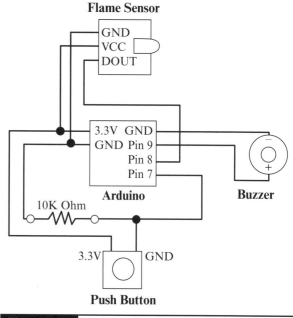

Figure 5-11 Arduino fire alarm.

- Connect the other terminal of the push button to a 10-kΩ resistor. Connect pin 7 of the Arduino to this node. Connect the other end of the resistor to ground (Figure 5-11).

Setting Up the Software

The program for this example project implements a basic fire alarm that activates a buzzer when a flame is detected and can be reset by pressing a button. The program builds on the preceding program and has added:

1. A variable called `AlarmTripped` that is true when the fire alarm has detected a flame and will stay true until reset. This variable is initialized to false.

2. A variable called `BuzzerPin` that represents the positive terminal of the buzzer, which is assigned to digital pin 9 of the Arduino.

3. A variable called `BuzzerFreq` that holds the frequency of the buzzer sound to be produced when the alarm is tripped and is initialized to 300.

4. A variable called `ButtonPin` that represents the button and is assigned to digital pin 7 of the Arduino.

5. The message displayed on the Serial Monitor indicating that the program has started is updated.

6. In the `loop()` function, the following modifications and additions were made:

    a. The button status is read, and if the button was pressed, then the alarm is reset.

    b. If the alarm is tripped, then the program produces a tone using the buzzer. It also prints a message to the Serial Monitor indicating that the alarm has been tripped.

    c. If the alarm is not tripped, then the program stops any sound being produced by the buzzer.

    d. If a flame has been detected, then the program trips the alarm by setting the `AlarmTripped` variable to true. See Listing 5-7. The new code is shown in bold type.

Listing 5-7   Arduino Fire Alarm

```
// Fire Alarm
int FlameOutPin = 8;
int RawValue = 0;
boolean AlarmTripped = false;

// Buzzer
int BuzzerPin = 9;
int BuzzerFreq = 300;

// Button
int ButtonPin = 7;

void setup()
{
  pinMode(FlameOutPin, INPUT);
  Serial.begin(9600);
  Serial.println("Fire Alarm ...");
```

```
}

void loop()
{
  // Read Reset Button
  RawValue = digitalRead(ButtonPin);
  if (RawValue == 1)
  {
      AlarmTripped = false;
  }

  // Check if Alarm is Tripped
  if(AlarmTripped)
  {
      tone(BuzzerPin, BuzzerFreq);
      Serial.print("Alarm Tripped ... ");
  }
  else
  {
      noTone(BuzzerPin);
  }

  // Read Flame Sensor
  RawValue = digitalRead(FlameOutPin);
  if (RawValue == 0)
  {
      Serial.print("Flame Detected ...");
      AlarmTripped = true;
  }
  else
  {
      Serial.print("No Flame Detected
        ...");
  }
  // Print out raw value from sensor
  Serial.print(" , RawValue: ");
  Serial.println(RawValue);
}
```

Running the Program

Upload the program to your Arduino, and start the Serial Monitor. The Serial Monitor will display the current status of the flame sensor. Use a flame or a flashlight to activate the flame sensor. You should hear the buzzer activate and stay on even when the flame or flashlight has been removed. Press the button to reset the alarm. The buzzer should deactivate, and the fire alarm system is now ready to detect the next occurrence of a flame.

Hands-on Example: Raspberry Pi Flame Sensor Test

In this hands-on example project, you will learn how to connect and use a flame sensor with a Raspberry Pi.

Parts List

The parts you will need to construct this example project include

- 1 flame sensor
- 1 breadboard (optional to hold flame sensor)
- Wires to connect the flame sensor to the Raspberry Pi

Setting Up the Hardware

To build this hands-on example project, you will need to:

- Connect the GND pin on the flame sensor to a GND pin of the Raspberry Pi.
- Connect the VCC pin on the flame sensor to a 3.3-V pin of the Raspberry Pi.
- Connect the DOUT pin on the flame sensor to pin 14 of the Raspberry Pi (Figure 5-12).

Setting Up the Software

The program for this example project tests the flame sensor by:

1. Importing the RPi.GPIO library.

2. Assigning pin 14 of the Raspberry Pi to the output pin on the flame sensor, which is represented by the `FlameOutPin` variable.

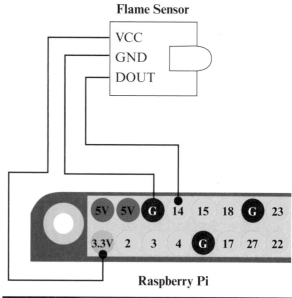

Figure 5-12   Raspberry Pi flame sensor test.

3. Initializing `RawValue`, which is the variable that reads data from the flame sensor to 0.

4. Setting the general-purpose input-output (GPIO) pin numbering system to BCM mode.

5. Setting the `FlameOutPin` as an input pin so that data from the flame sensor can be read.

6. Printing a message to the terminal indicating that the program has started running.

7. Doing the following until a keyboard interrupt has been detected:

    a. Read the value from the flame sensor output pin.

    b. If the value read is 0, then a flame has been detected, so print out a message to the terminal indicating that a flame was found.

    c. If the value read is not 0, then no flame is detected, so print out a message to the terminal indicating that no flame was found.

8. Print out the raw value read from the flame sensor in step 1.

9. Printing out a message indicating that the program is terminating.

10. Deallocating GPIO pin resources by calling the `GPIO.cleanup()` function. See Listing 5-8.

Listing 5-8   Raspberry Pi Flame Sensor Test

```
# Flame Sensor Test
import RPi.GPIO as GPIO

FlameOutPin = 14
RawValue = 0

# Setup
GPIO.setmode(GPIO.BCM)
GPIO.setup(FlameOutPin, GPIO.IN)
print("Flame Sensor Test ...")

try:
  while 1:
     RawValue = GPIO.input(FlameOutPin)
     if (RawValue == 0):
        print("Flame Detected !!!!!")
     else:
        print("No Flame Detected ....")

     print(" , RawValue: ", RawValue)

except KeyboardInterrupt:
 pass
print("Exiting Flame Sensor Test ...")
GPIO.cleanup()
```

Running the Program

Start the program from the Raspberry Pi terminal using Python. You should see the flame detection status displayed on the terminal. Use a flame from a match or lighter or light from a flashlight to activate the flame sensor. You should see the status of the flame detector change on the monitor.

Hands-on Example: Raspberry Pi Fire Alarm

In this hands-on example project, you will learn how to build a Raspberry Pi–based fire alarm. The fire alarm uses the flame sensor from the preceding hands-on example project. Once a flame is detected, an alarm sound plays until the system is reset by the user by pressing the reset button.

Parts List

To construct this example project, you will need

- 1 flame sensor
- 1 push button
- 1 10-kΩ resistor
- 1 breadboard (optional to hold the flame sensor)
- Wires to connect the components together

Setting Up the Hardware

To set up the hardware for this example project, you will need to:

- Connect the GND pin on the flame sensor to a GND pin of the Raspberry Pi.
- Connect the VCC pin on the flame sensor to the 3.3-V pin of the Raspberry Pi.
- Connect the DOUT pin on the flame sensor to pin 14 of the Raspberry Pi.
- Connect one terminal on the push button to the 3.3-V pin of the Raspberry Pi.
- Connect the other terminal on the push button to a node containing a 10-kΩ resistor. Connect pin 15 of the Raspberry Pi to this node. And connect the other end of the resistor to a GND pin of the Raspberry Pi (Figure 5-13).

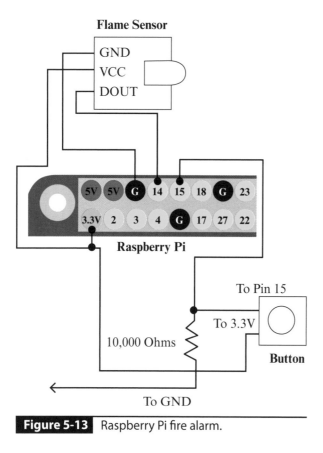

Figure 5-13 Raspberry Pi fire alarm.

Setting Up the Software

The program in the example project detects a flame and activates an alarm that is played though the speaker on the monitor attached to the Raspberry Pi. The alarm can be reset by pressing the push button attached to the GPIO pins on the Raspberry Pi. The program:

1. Imports the RPi.GPIO library so that it can be used by the program.

2. Imports the pygame library so that sound effects can be used by the program.

3. Assigns the FlameOutPin, which represents the output pin on the flame detector, to pin 14 of the Raspberry Pi.

4. Initializes to 0 the RawValue variable, which holds the value from the flame sensor read by the Raspberry Pi.

5. Initializes to 0 the AlarmTripped variable, which is 0 if the alarm has not detected a flame and 1 if the alarm has detected a flame.

6. Assigns pin 15 of the Raspberry Pi to the ButtonPin variable, which represents the button the user presses to reset the alarm.

7. Sets the pin numbering method on the Raspberry Pi to BCM mode.

8. Sets the FlameOutPin pin on the Raspberry Pi to be an input pin that can read the voltage at that pin. This pin is connected to the flame detector output pin.

9. Sets the ButtonPin pin on the Raspberry Pi to be an input pin that can read voltages. This pin is connected to the alarm reset button.

10. Prints a text message to the terminal to indicate that the program has started to run.

11. Initializes the pygame library and sound system.

12. Creates a sound effect object called Alarm from a .wav sound file.

13. Does the following until the user generates a keyboard interrupt by pressing CTRL-C:

    a. If the user presses the reset button, resets the alarm by setting the AlarmTripped variable to 0.

    b. If the alarm has been tripped, plays the Alarm sound effect and prints a text message to the terminal indicating that the alarm has been tripped.

    c. If the alarm has not been tripped, halts the playing of the Alarm sound effect.

    d. Reads the flame detector sensor. If a flame has been detected, trips the alarm and prints a message to the terminal indicating that a flame has been detected. If a flame has not been detected, prints a message to the terminal indicating that no flame has been detected.

e. Prints the raw value read from the flame detector.

14. Prints a text message to the terminal indicating that the program is exiting.

15. Deallocates resources related to the GPIO pins. See Listing 5-9.

Listing 5-9   Raspberry Pi Fire Alarm

```
# Fire Alarm
import RPi.GPIO as GPIO
import pygame

FlameOutPin = 14
RawValue = 0
AlarmTripped = 0

# Button
ButtonPin = 15

#Setup
GPIO.setmode(GPIO.BCM)
GPIO.setup(FlameOutPin, GPIO.IN)
GPIO.setup(ButtonPin, GPIO.IN)
print("Fire Alarm ...")

#SFX
pygame.init()
pygame.mixer.init()
Alarm = pygame.mixer.Sound("match1.wav")

try:
 while 1:
    # Read Reset Button
    RawValue = GPIO.input(ButtonPin)
    if (RawValue == 1):
       AlarmTripped = 0

    # Check if Alarm is Tripped
    if(AlarmTripped):
       Alarm.play()
       print("Alarm Tripped ... ")
    else:
       Alarm.stop()

    # Read Flame Sensor
    RawValue = GPIO.input(FlameOutPin)
```

```
    if (RawValue == 0):
       print("Flame Detected ...")
       AlarmTripped = 1
    else:
       print("No Flame Detected ...")

    # Print out raw value from sensor
    print(" , RawValue: ", RawValue)

except KeyboardInterrupt:
 pass
print("Exiting PI Fire Alarm ...")
GPIO.cleanup()
```

Running the Program

Run the program from the terminal using Python. On the terminal you should see the current status of the fire alarm, which should not be tripped and not detecting a flame. Next, activate the flame sensor by using a flashlight or flame. You should hear a sound from the speaker of your monitor. This sound should continue even if you remove the light or flame from the flame detector. This fire alarm has now been tripped and will stay stripped until you press the reset button. Press the reset button with your finger. The sound should stop. The fire alarm has been reset and is ready to detect another occurrence of a flame.

Infrared Proximity/Collision Sensor

An infrared proximity/collision sensor detects how close the sensor at the end of the board is to another object using infrared radiation (Figure 5-14).

Figure 5-14   Infrared proximity/collision sensor.

Hands-on Example: Arduino Infrared Proximity/Collision Detector

In this hands-on example project, you will learn how to build an infrared proximity/collision detector using an Arduino and a piezo buzzer. The buzzer will produce a tone whenever an object is brought in close proximity to the sensor.

Parts List

To build this example project, you will need

- 1 infrared proximity/collision sensor
- 1 piezo buzzer
- 1 breadboard (optional to hold the sensor)
- Wires to connect the components together

Setting Up the Hardware

To set up the hardware for this example, you will need to:

- Connect the GND pin on the sensor to a GND pin of the Arduino.

- Connect the + or VCC pin on the sensor to the 3.3-V pin of the Arduino.

- Connect the Out pin on the sensor to digital pin 8 of the Arduino.

- Connect the negative terminal on the piezo buzzer to a GND pin of the Arduino.

- Connect the positive terminal on the piezo buzzer to digital pin 9 of the Arduino (Figure 5-15).

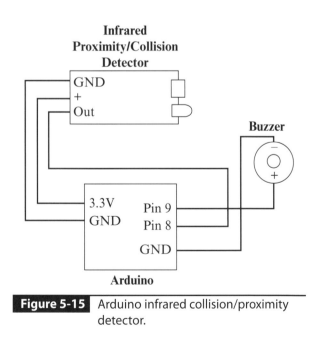

Figure 5-15   Arduino infrared collision/proximity detector.

Setting Up the Software

The program for this hands-on example project detects the closeness of an object to the sensor and produces a tone via a piezo buzzer when this occurs by:

1. Assigning pin 8 of the Arduino to the output pin on the sensor, which is represented by the IRSensorPin variable.

2. Initializing RawValue to 0. This variable holds the data read from the IR proximity sensor.

3. Assigning pin 9 of the Arduino to BuzzerPin, which represents the positive terminal of the piezo buzzer.

4. Initializing to 300 the BuzzerFreq variable, which represents the frequency of the tone generated by the buzzer.

5. Defining the setup() function, which:

   a. Sets the IRSensorPin variable, which represents the pin on the Arduino that is connected to the IR proximity sensor, to be an input pin, which means that the voltages at the pin can be measured.

   b. Initializes the Serial Monitor to 9,600 baud.

   c. Prints a text message to the Serial Monitor indicating that program execution has started.

6. Defines the loop() function, which:

   a. Reads the proximity collision status from the sensor.

   b. If the status is 0, an object has been detected near the sensor, so prints a message to the terminal indicating that a pending collision has been detected and starts generating a sound with the buzzer.

   c. If the status is not 0, no object has been detected near the sensor, so prints a message to the terminal indicating that no object has been detected and stops playing any sound via the buzzer.

   d. Prints the raw value read from the sensor to the terminal. See Listing 5-10.

Listing 5-10 Arduino Infrared Collision/Proximity Detector

```
// Infrared Proximity Sensor
int IRSensorPin = 8;
int RawValue = 0;

// Buzzer
int BuzzerPin = 9;
int BuzzerFreq = 300;

void setup()
{
 pinMode(IRSensorPin, INPUT);
 Serial.begin(9600);
 Serial.println("Arduino Infrared Proximity Sensor ...");
}

void loop()
```

(continued on next page)

Listing 5-10 Arduino Infrared Collision/Proximity Detector (*continued*)

```
{
 RawValue = digitalRead(IRSensorPin);
 if(RawValue == 0)
 {
    Serial.print("IR Proximity Sensor Activated ...");
    tone(BuzzerPin, BuzzerFreq);
 }
 else
 {
    Serial.print("No Object Detected ...");
    noTone(BuzzerPin);
 }
 Serial.print(" , RawValue: ");
 Serial.println(RawValue);
}
```

Running the Program

Upload the program to your Arduino, and start the Serial Monitor. You should see the status of the object detection displayed on the terminal window, which should be that no object is detected. Bring a flat piece of white paper within an inch or less of the sensor. The buzzer should sound, and the status displayed on the terminal screen should change to indicate that an object has been detected. Note that the best object to use for this type of test is a reflective, flat piece of white paper.

Hands-on Example: Raspberry Pi Infrared Collision/Proximity Sensor Alarm

In this hands-on example project, you will learn how to connect an infrared proximity/collision detector sensor with a Raspberry Pi. When an object is detected by the sensor, the Raspberry Pi will produce a sound effect.

Parts List

To build this example project, you will need

- 1 infrared collision/proximity sensor
- 1 breadboard (optional in order to hold the sensor)
- Wires to connect the sensor to the Raspberry Pi

Setting Up the Hardware

To build the hardware in this example project, you will need to:

- Connect the GND pin on the infrared proximity sensor to a GND pin of the Raspberry Pi.
- Connect the + or VCC pin on the sensor to a 3.3-V pin of the Raspberry Pi.
- Connect the output pin on the sensor to pin 14 of the Raspberry Pi (Figure 5-16).

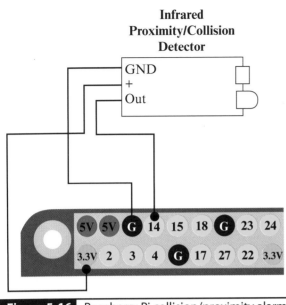

**Infrared
Proximity/Collision
Detector**

GND
+
Out

Figure 5-16 Raspberry Pi collision/proximity alarm.

Setting Up the Software

The program for this example project produces a sound when an object is close to the infrared sensor by:

1. Importing the GPIO library into the program so that it can be used.

2. Importing the pygame library into the program so that it can be used.

3. Assigning pin 14 of the Raspberry Pi to the `IRSensorPin` variable, which represents the output sensor pin on the infrared proximity detector.

4. Initializing the `RawValue` variable to 0. This variable is used to read data from the sensor.

5. Setting the Raspberry Pi pin numbering method to BCM mode.

6. Setting the `IRSensorPin` pin to be an input pin, which allows the Raspberry Pi to measure the voltage level at that pin.

7. Printing a text message to the terminal indicating that the program has started.

8. Initializing the pygame library and the pygame sound effect system.

9. Creating a sound effect object called `Alarm` from a .wav sound file that is located in the Raspberry Pi's file system.

10. Doing the following until a keyboard interrupt is generated by the user:

    a. Read the value output from the sensor.

    b. If the value is 0, an object has been detected, so print a notification to the terminal and start playing a sound effect.

    c. If the value is not 0, no object has been detected, so print a text message to the terminal indicating that no object has been detected and stop playing the sound effect.

    d. Print the raw value read from the sensor to the terminal.

11. Printing a text message to the terminal indicating that the program is ending.

12. Deallocating GPIO resources. See Listing 5-11.

Listing 5-11   Raspberry Pi Collision/Proximity Alarm

```
# Infrared Proximity Sensor
import RPi.GPIO as GPIO
import pygame


IRSensorPin = 14
RawValue = 0


#Setup
GPIO.setmode(GPIO.BCM)
GPIO.setup(IRSensorPin, GPIO.IN)
print("Arduino Infrared Proximity Sensor ...")


#SFX
pygame.init()
pygame.mixer.init()
Alarm = pygame.mixer.Sound("match1.wav")


try:
 while 1:
    RawValue = GPIO.input(IRSensorPin)
    if(RawValue == 0):
       print("IR Proximity Sensor Activated ...")
       Alarm.play()
    else:
       print("No Object Detected ...")
       Alarm.stop()

    print(" , RawValue: ", RawValue)
except KeyboardInterrupt:
 pass
print("Exiting IR Sensor Collision Detect ...")
GPIO.cleanup()
```

Running the Program

Before starting the program, make sure that you have a sound file in the same directory as the script you are executing. If needed, change the sound filename in the script to reflect the sound file you will be using. Start the program using Python. The output on the terminal should indicate that no object has been detected. Place a flat piece of white paper an inch or less in front of the sensor. The alarm should sound, and the message on the terminal should indicate that

an object as been detected. Pull the paper away from the sensor, and the sound should stop and the text message on the screen should indicate that no object has been detected.

Temperature and Humidity Sensor (DHT11)

The DHT11 is a digital sensor that measures temperature and humidity (Figure 5-17).

Figure 5-17 DHT11 sensor.

Operational Information

The DHT11 sensor can operate on both 3.3 and 5 volts. The data from the sensor consist of 40 bits, or 5 bytes, of continuous data. The data consist of the integer part of the relative humidity, followed by the decimal part of the relative humidity, followed by the integer part of the temperature in Celsius, followed by the decimal part of the temperature, and finally, the checksum. The checksum is the last 8 bits of the sum of all the previous 4 bytes of data (Figure 5-18).

In addition, the data come in with the most significant bit from each byte first. So a good graphic representation of the incoming data stream would be Figure 5-19.

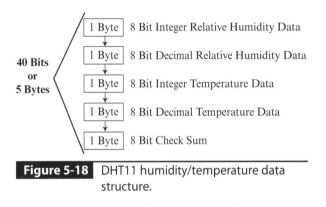

Figure 5-18 DHT11 humidity/temperature data structure.

Requesting Data from the DHT11 Sensor

To retrieve temperature and humidity data from the DHT11 sensor, you will need to:

1. Send a request to retrieve data from the DHT11. Have your microcontroller such

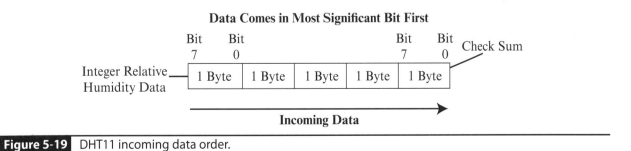

Figure 5-19 DHT11 incoming data order.

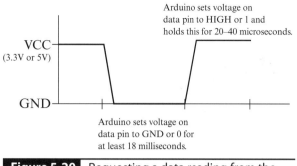

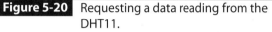

Figure 5-20 Requesting a data reading from the DHT11.

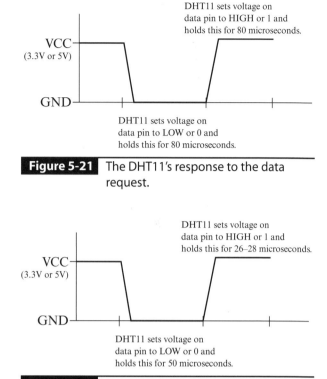

Figure 5-21 The DHT11's response to the data request.

Figure 5-22 The DHT11 sending a 0 data bit.

as your Arduino set the voltage on the sensor's data pin to GND or 0 for at least 18 milliseconds. Then set the voltage on the data pin to HIGH or 1, and hold this for 20 to 40 microseconds. This will signal the DHT11 sensor that it needs to provide temperature and humidity data (Figure 5-20).

2. Read the response from the DHT11 sensor to make sure that the DHT11 has acknowledged the data request. The DHT11 sets the voltage LOW on the data pin for 80 microseconds and then sets the voltage HIGH on the data pin for 80 microseconds (Figure 5-21).

3. Read in the data, which are 5 bytes total, one bit at a time. A 0 bit is indicated by the DH11 setting the voltage LOW on

the data pin and holding it there for 50 microseconds, followed by setting the voltage HIGH and holding it there for 26 to 28 microseconds (Figure 5-22).

4. A 1 data bit is generated by the DHT11 by setting the voltage LOW on the data pin for 50 microseconds and then setting the voltage HIGH on the data pin for 70 microseconds (Figure 5-23).

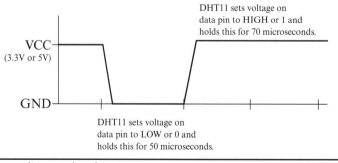

Figure 5-23 The DHT11 sending a 1 data bit.

Hands-on Example: Arduino DHT11 Temperature and Humidity Sensor Test

In this hands-on example project, you will learn how to connect a DHT11 temperature and humidity sensor to an Arduino.

Parts List

The components you will need to build this example project include

- 1 DHT11 temperature and humidity sensor
- 1 breadboard (optional to hold the sensor)
- Wires to connect the components together

Setting Up the Hardware

To build the hardware for this example project, you will need to:

- Connect the GND pin on the DHT11 to a GND pin of the Arduino.
- Connect the + or VCC pin on the DHT11 to the 3.3-V pin of the Arduino.
- Connect the DAT pin on the DHT11 to digital pin 8 of the Arduino (Figure 5-24).

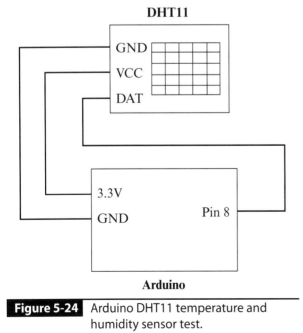

Figure 5-24 Arduino DHT11 temperature and humidity sensor test.

Setting Up the Software

The program for this example project continuously retrieves data from the DHT11 sensor and prints the humidity and temperature to the Serial Monitor. Pin 8 of the Arduino is assigned to the data pin on the DHT11 by the initialization statement

```
int DHT11Pin = 8;
```

The RawValue variable, which holds the incoming data from the DHT11, is initialized to 0:

```
int RawValue = 0;
```

The number of bytes of an incoming temperature/humidity reading is set to 5:

```
const int NumEntries = 5;
```

The total number of bits in an incoming temperature/humidity reading is set to 40:

```
const int TotalDataBits = 40;
```

The incoming data from the DHT11 are held in the Data array, with each element in the array representing 1 byte from the temperature/humidity data:

```
byte Data[NumEntries];
```

The setup() function is executed first by the Arduino, and the function:

1. Initializes the Serial Monitor with a communication speed of 9,600 baud.
2. Prints a text message to the Serial Monitor to indicate that the program has started running.
3. Suspends program execution for 1.5 seconds to allow the DHT11 to stabilize before issuing any commands to it. See Listing 5-12.

Listing 5-12 The `setup()` Function

```
void setup()
{
 Serial.begin(9600);
 Serial.println("DHT11 Test ...");
 delay(1500);
}
```

The `loop()` function is continuously executed by the Arduino and is where the temperature and humidity data from the DHT11 are continuously read in and printed out to the Serial Monitor by:

1. Resetting all the bytes of data in the `Data` array to 0 by calling the `ResetData()` function.

2. Sending a request for a temperature/humidity data reading to the DHT11 sensor from the Arduino by calling the `StartSignalToDHT11()` function.

3. Reading the response signal from the DHT11 that indicates that the DHT11 has received the Arduino's request and is ready to send the data by calling the `ReadDHT11StartSignal()` function.

4. If the return value from step 3 is false, then printing out an error message to the Serial Monitor indicating that the DHT11's response to the Arduino's start signal failed.

5. If the return value from step 3 is true, then reading the 5-byte or 40-bit data entry from the DHT11 that holds the temperature and humidity data by calling the `ReadDataEntryDHT11()` function.

6. If the return value from step 5 is false, then printing a text message to the Serial Monitor that there was an error in the checksum. Otherwise, printing a text message saying that the checksum was okay.

7. Printing the temperature, humidity, checksum, and checksum status to the Serial Monitor.

8. Suspending execution of the program for 1,000 milliseconds, or 1 second, by calling the `delay()` function with 1,000 as the input parameter. See Listing 5-13.

Listing 5-13 The `loop()` Function

```
void loop()
{
 boolean result = false;
 // Reset data structures that will hold incoming data from the DHT11 to 0.
 ResetData();

 // Send start signal to the DHT11 sensor
 StartSignalToDHT11();

 // Read in the DHT response to the start signal to indicate that the sensor is
 // ready to send data
 result = ReadDHT11StartSignal();
 if (result == false)
 {
     Serial.println("DHT11 START SIGNAL RESPONSE FAILURE ....");
 }

 // Read in 40 bits or the 5 byte data entry sent by the DHT that represents the
```

```
// relative humidity and the temperature.
result = ReadDataEntryDHT11();
if (result == false)
{
   Serial.print("CheckSum Error .... ");
}
else
{
   Serial.print("CheckSum Ok ... ");
}
// Print Data Recieved from the DHT11 sensor
PrintData();

// Delay next reading
delay(1000);
}
```

The `ResetData()` function sets all of the elements in the `Data` array to 0. See Listing 5-14.

Listing 5-14 The `ResetData()` function

```
void ResetData()
{
 for(int i = 0; i < NumEntries; i++)
 {
    Data[i] = 0;
 }
}
```

The `StartSignalToDHT11()` function requests data from the DHT11 sensor by:

1. Setting the pin on the Arduino that is connected to the data pin on the DHT11 sensor to be an output pin so that a voltage signal can be output to the DHT11. This is done using the `pinMode()` function.

2. Setting the voltage on the pin connected to the data pin on the DHT11 to `LOW` or 0 using the `digitalWrite()` function.

3. Holding the `LOW` voltage value on the pin by suspending program execution for 18 milliseconds by calling the `delay()` function.

4. Setting the voltage on the data pin to `HIGH` by calling the `digitalWrite()` function.

5. Holding the voltage on the data pin `HIGH` by suspending execution of the program for 40 microseconds. See Listing 5-15.

Listing 5-15 The `StartSignalToDHT11()` Function

```
void StartSignalToDHT11()
{
 pinMode(DHT11Pin, OUTPUT);
 digitalWrite(DHT11Pin, LOW);
 delay(18);
 digitalWrite(DHT11Pin, HIGH);
 delayMicroseconds(40);
}
```

The `ReadDHT11StartSignal()` function reads the DHT11 response to the start signal sent by the Arduino by:

1. Setting the Arduino pin connected to the data pin on the DHT11 as an input pin that will read the voltages generated from the DHT11.

2. Saving the current time since startup in microseconds.

3. While the value on the DHT11 data pin is LOW or 0, do the following:

   a. Read the value of the data pin.

   b. Calculate the elapsed time since the start of the while() loop.

   c. If the value read from the data pin is 1 and the elapsed time is greater than 70 microseconds, then exit the loop.

   d. If the value read in from the data pin is 0 and the elapsed time is greater than 90 microseconds, then exit the function by returning false, which indicates an error condition.

4. Saving the current time since start-up in microseconds.

5. While the value on the DHT11 data pin is HIGH or 1, do the following:

   a. Read the value of the data pin.

   b. Calculate the elapsed time since the start of the while() loop.

   c. If the value of the data pin is LOW or 0 and the elapsed time is greater than 70 microseconds, then exit the while() loop.

   d. If the value of the data pin is HIGH or 1 and the elapsed time is greater than 90 microseconds, then return false, which indicates an error condition.

6. Returning true if the DHT11 responded correctly or false if there was an error condition. See Listing 5-16.

Listing 5-16 The ReadDHT11StartSignal() Function

```
boolean ReadDHT11StartSignal()
{
  // Returns true if DHT sends a low voltage for 80us and
  // then a high voltage for 80us then goes low.
  // Returns false otherwise.

  boolean result = true;
  boolean done = false;
  unsigned long StartTime = 0;
  unsigned long CurrentTime = 0;
  unsigned long ElapsedTime = 0;

  // Change pin to read in data from DHT11
  pinMode(DHT11Pin, INPUT);

  // Wait until value goes from low to high after 80us or times out
  StartTime = micros();
  done = false;
  while(!done)
  {
    // Read in pin value
    RawValue = digitalRead(DHT11Pin);
    CurrentTime = micros();
    ElapsedTime = CurrentTime - StartTime;
    if (RawValue == 1)
    {
```

```
      if (ElapsedTime > 70)
      {
         done = true;
      }
   }
   else
   {
      // Check Timeout
      if(ElapsedTime > 90)
      {
         // Response Takes Too Long
         return false;
      }
   }
}

// Wait until value goes from high for 80us to low or times out.
StartTime = micros();
done = false;
while(!done)
{
   // Read in pin value
   RawValue = digitalRead(DHT11Pin);
   CurrentTime = micros();
   ElapsedTime = CurrentTime - StartTime;
   if (RawValue == 0)
   {
      if (ElapsedTime > 70)
      {
         done = true;
      }
   }
   else
   {
      // Check Timeout
      if(ElapsedTime > 90)
      {
         // Response Takes Too Long
         return false;
      }
   }
}
return result;
}
```

The `ReadDataEntryDHT11()` function reads in the 40 bits, or 5 bytes, of incoming data from the DHT11 by doing the following for each of the 40 incoming bits:

1. Reading the data from the DHT11 sensor's data pin until a value of 1 or `HIGH` is found. This indicates that the next bit will be an actual data bit.

2. Saving the current time since power-up in microseconds.

3. While the value on the DHT11's data pin is 1 or `HIGH`, do the following:

   a. Read the value of the DHT11's data pin.

   b. If the value is 0, Find the elapsed time since the start of the `while()` loop.

   c. If the elapsed time is greater than 60 microseconds, then the incoming bit is a 1, so put a 1 in the appropriate bit position in the `Data` array based on the `BitPos` and `BytePos` variables. Remember, there are 8 bits in a byte and 5 bytes total in the data we are reading. The data are coming with the most significant bit first (bit 7) for each byte.

   d. If the current `BitPos` is 0, the current byte has been read in, and we need to go to the next byte. To do this, we reset the `BitPos` to 7, which is bit 7, and increase the value of BytePos to indicate that we will be reading data for the next byte.

   e. If the current `BitPos` is not 0, decrease `BitPos` by 1.

4. Calculating the checksum as the last 8 bits of the sum of the first 4 bytes of the data that were just read.

5. If the checksum calculated in step 4 is equal to the checksum byte, which was the final byte read, then the data have been validated. If the calculated checksum does not match the checksum byte, then the incoming data were corrupted.

6. Returning the checksum validation status to the calling function. See Listing 5-17.

Listing 5-17 The `ReadDataEntryDHT11()` Function

```
boolean ReadDataEntryDHT11()
{
  boolean result = false;
  byte CheckSum = 0;
  int BytePos = 0;
  int BitPos = 7;
  boolean done = false;
  unsigned long StartTime = 0;
  unsigned long CurrentTime = 0;
  unsigned long ElapsedTime = 0;
  byte temp = 0;

  // Read in all 5 bytes of data 1 bit at a time
  // Most significant bit is read in first.
  for (int i = 0; i < TotalDataBits; i++)
  {
      // Wait for data bit
```

```
   done = false;
   while(!done)
   {
      RawValue = digitalRead(DHT11Pin);
      if (RawValue == 1)
      {
         done = true;
      }
   }

   // Read in data bit
   done = false;
   StartTime = micros();
   while(!done)
   {
      RawValue = digitalRead(DHT11Pin);
      if (RawValue == 0)
      {
         CurrentTime = micros();
         ElapsedTime = CurrentTime - StartTime;
         if (ElapsedTime > 60)
         {
            // Data is 1
            temp = (1 << BitPos);
            Data[BytePos] = Data[BytePos] | temp;
         }

         // Update Bit and Byte Positions
         if(BitPos == 0)
         {
            BitPos = 7;
            BytePos++;
         }
         else
         {
            BitPos--;
         }
         done = true;
      }
   }
}

// Calculate Checksum
CheckSum = Data[0] + Data[1] + Data[2] + Data[3];
if (CheckSum == Data[4])
{
   result = true;
}
return result;
}
```

The `PrintData()` function prints key data to the Serial Monitor by:

1. Printing the humidity read from the DHT11 sensor.

2. Printing the temperature in Celsius read from the DHT11 sensor.

3. Printing the temperature in Fahrenheit.

4. 4. Printing the checksum read from the DHT11 sensor. See Listing 5-18.

Listing 5-18 The `PrintData()` Function

```
void PrintData()
{
 Serial.print("Humidity: ");
 Serial.print(Data[0]);
 Serial.print(".");
 Serial.print(Data[1]);

 Serial.print(" , TempC: ");
 Serial.print(Data[2]);
 Serial.print(".");
 Serial.print(Data[3]);

 Serial.print(" , TempF: ");
 Serial.print(GetTempFarenheit(Data[2]));

 Serial.print(" , CheckSum: ");
 Serial.println(Data[4]);
}
```

The `GetTempFarenheit()` function converts a Celsius temperature to a Fahrenheit temperature and returns the result. See Listing 5-19.

Listing 5-19 The `GetTempFarenheit()` Function

```
float GetTempFarenheit(float TempC)
{
 float temp = 0;
 temp = TempC *(9.0/5.0) + 32.0;
 return temp;
}
```

Running the Program

Upload the program to your Arduino, and start the Serial Monitor. You should see the checksum status, the humidity, the temperature in Celsius, the temperature in Fahrenheit, and the checksum values printed in the Serial Monitor similar to:

```
CheckSum Ok ... Humidity: 38.0 , TempC:
22.0 , TempF: 71.60 , CheckSum: 60
```

These values are continually read from the DHT11 sensor and printed out.

Summary

This chapter covered environmental sensors for the Arduino and the Raspberry Pi. First, the reed switch was discussed. Hands-on example projects using this sensor were illustrated for the Arduino and the Raspberry Pi, including a door entry alarm system. Next, the TMP36 analog temperature sensor was introduced and used with the Arduino. The flame sensor came next and was used in systems such as a fire alarm system for both the Arduino and Raspberry Pi. Next, we introduced the infrared proximity/collision detector and provided collision alarm examples for both the Arduino and Raspberry Pi. The chapter concluded with the DHT11 temperature and humidity sensor and an example project with the Arduino.

## CHAPTER 6

Environmental Sensor Projects II

THIS CHAPTER DISCUSSES some important environmental sensors. We begin by covering a water detector or soil moisture sensor. Hands-on examples employing this sensor as a water or flood detector alarm and soil moisture sensor with the Arduino are presented. Next is a discussion of a light detector, or what is also known as a photo resistor. In this section you learn how to make a "rooster alarm" that activates when the sun rises. A sound detector sensor is introduced next. Here you use this sensor in an Arduino project that allows you to clap two times to turn a LED on and off. You will also use this sensor in a Raspberry Pi game called "Out of Breath," where players use their voices to play the game.

Water Detector/Soil Moisture Sensor

A water detector or soil moisture sensor detects the amount of water or moisture in the environment. One good use for this device is to detect water leaks from a device such as a hot water heater before they become critical. Another good use for this device is to measure the moisture in soil so that you have a good idea when to water your plants. The device itself consists of two parts. One part has connections to the microcontroller and to the part of the device that does the actual measurement (Figure 6-1).

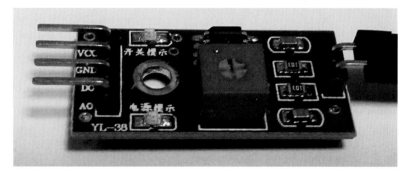

Figure 6-1 Water detector component connected to an Arduino

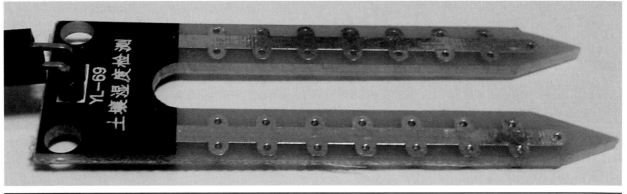

Figure 6-2   The water detector component that measures the amount of moisture.

The part of the water detector that actually measures the moisture in the environment is a forklike device, as shown in Figure 6-2.

Hands-on Example: Arduino Water Detector Alarm

In this hands-on example project, you will learn how to build and operate a water detector alarm that is appropriate as an alarm to detect water around a hot water heater or other area of your home. A hot water heater is used in your home to produce the hot water that flows from a water spigot when you turn on the "hot" knob of a sink. Most hot water heaters are basically huge tanks of hot water, and eventually, all of these tanks will fail and start to leak. If the leak is not caught early, then you can have a big problem.

Parts List

For this hands-on example project, you will need

- 1 water detector
- 1 piezo buzzer
- 1 push button
- 1 10-kΩ resistor (color coded brown, black, and orange)
- 1 breadboard (optional to hold the water detector)

- Wires to connect all the components to the Arduino

Setting Up the Hardware

To construct the hardware for this example project, you will need to:

- Connect the VCC pin on the water detector to the 5-V pin of the Arduino.
- Connect the GND pin on the water detector to the GND pin of the Arduino.
- Connect the AO pin on the water detector to analog pin A0 of the Arduino.
- Connect the GND or negative terminal on the piezo buzzer to a GND pin of the Arduino.
- Connect the positive terminal on the piezo buzzer to digital pin 9 of the Arduino.
- Connect one terminal on the push button to the 5-V pin of the Arduino.
- Connect the other terminal on the push button to a node that contains a 10-kΩ resistor. Connect this node to digital pin 7 of the Arduino.
- Connect the other end of the resistor to a GND pin of the Arduino (Figure 6-3).

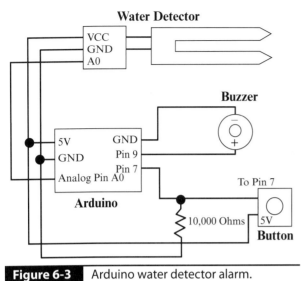

Water Detector

VCC
GND
A0

Buzzer

5V   GND
GND   Pin 9
          Pin 7
Analog Pin A0

Arduino

To Pin 7

10,000 Ohms   5V

Button

 Arduino water detector alarm.

Setting Up the Software

A water detector sensor measures the moisture level in the environment. A low value from the water detector means that the environment is very moist. A high value from the water detector means that the environment is very dry. Basically, the sensor measures the voltage across the two prongs of the forklike part of the sensor. When the voltage is high, the resistance is high because there is no substance like water between the prongs. When the voltage is low, the resistance is low because there is a substance that conducts electricity (like water) between the prongs. Recall from Chapter 3 that voltage = resistance × current.

Experimentally, I determined some values from the water detector that correspond to various amounts of wetness or dryness. If the environment is very wet, then the value from the water detector will be in the range of 0 to 400:

```
const int VERY_WET = 400;
```

If the environment is wet, then the value from the water detector will be in the range of 401 to 500:

```
const int WET = 500;
```

If the environment is just damp, the value from the water detector will be in the range of 501 to 700:

```
const int DAMP = 700;
```

If the environment is dry, the value from the water detector will be in the range of 701 to 850:

```
const int DRY = 850;
```

If the environment is very dry, then the value from the water detector will be in the range of 851 to 950:

```
const int VERY_DRY = 950;
```

The piezo buzzer is assigned to digital pin 9 of the Arduino:

```
int BuzzerPin = 9;
```

The frequency of the sound the buzzer produces is set to 300:

```
int BuzzerFreq = 300;
```

The push button is assigned to digital pin 7 of the Arduino:

```
int ButtonPin = 7;
```

The `AlarmTripped` variable is true if the water detector has sensed a damp environment and false otherwise. This variable is initialized to false:

```
boolean AlarmTripped = false;
```

The water detector is assigned to analog pin 0 of the Arduino:

```
int DetectorPin = A0;
```

The `RawValue` variable holds the raw data that are read from the water detector and is initialized to 0:

```
int RawValue = 0;
```

The `setup()` function is called when the Arduino is powered on or reset and does the following:

1. The pin on the Arduino connected to the water detector's analog output pin is set as an input pin so that data can be read from the sensor.

2. The pin on the Arduino connected to the push button is set as an input pin so that voltages can be read to determine whether the button is being pushed.

3. The Serial Monitor is initialized, and the communication speed is set to 9,600 baud.

4. A text message is printed to the Serial Monitor indicating the start of the program. See Listing 6-1.

Listing 6-1 The setup() Function

```
void setup()
{
 pinMode(DetectorPin, INPUT);
 pinMode(ButtonPin, INPUT);
 Serial.begin(9600);
 Serial.println("Water Detector ...");
}
```

The PrintMoistureStatus() function prints out the moisture status of the water detector to the Serial Monitor, and the status is

1. VERY_WET,

2. WET,

3. DAMP,

4. DRY, or

5. VERY_DRY.

See Listing 6-2.

The loop() function is executed repeated until the Arduino is shut off or is reset, and it:

1. Reads the status of the push button and resets the alarm if the button is being pressed.

2. Reads the value from the water detector.

3. Prints out the level of moisture in the soil to the Serial Monitor based on the value from step 2 by calling the PrintMoistureStatus() function.

Listing 6-2 The PrintMoistureStatus() Function

```
void PrintMoistureStatus(int value)
{
 Serial.print("MoistureStatus: ");
 if (value <= VERY_WET)
 {
    Serial.print("VERY_WET");
 }
 else
 if (value <= WET)
 {
    Serial.print("WET");
 }
 else
 if (value <= DAMP)
 {
    Serial.print("DAMP");
 }
 else
 if (value <= DRY)
 {
    Serial.print("DRY");
 }
 else
 {
    Serial.print("VERY_DRY");
 }
}
```

4. Prints out the raw value read from the water detector to the Serial Monitor.

5. If the value read from the water detector from step 2 indicates that the moisture level is damp (which is 700 or less), then the function trips the alarm.

6. If the alarm has been tripped, the function produces a sound with the buzzer.

7. If the alarm has *not* been tripped, the function stops any sound being made by the buzzer. See Listing 6-3.

Listing 6-3   The `loop()` Function

```
void loop()
{
 // Read Reset Button
 RawValue = digitalRead(ButtonPin);
 if (RawValue == 1)
 {
    AlarmTripped = false;
 }

 RawValue = analogRead(DetectorPin);
 PrintMoistureStatus(RawValue);
 Serial.print(" , RawValue: ");
 Serial.println(RawValue);

 // Check to see if Alarm should be
 // tripped.
 if (RawValue <= DAMP)
 {
    AlarmTripped = true;
 }

 // Sound Alarm if alarm has been
 // tripped.
 if (AlarmTripped)
 {
    tone(BuzzerPin, BuzzerFreq);
 }
 else
 {
    noTone(BuzzerPin);
 }
}
```

Running the Program

Upload the program to the Arduino, and start the Serial Monitor. You should see output on the Serial Monitor similar to:

```
MoistureStatus: VERY_DRY , RawValue: 981
```

Because air is the substance between the two prongs in the sensor, the result is a high measurement of resistance, which translates into a "very dry" reading. Soak a paper towel with water and wrap it around the fork part of the water detector sensor. You should see the

reading go to anywhere from damp to very wet depending on how hard you press down on the fork with the wet towel. You should also activate the buzzer alarm when the reading gets to damp. The alarm is now tripped, and the buzzer will continuously produce a sound until you press the button to reset the alarm system.

Hands-on Example: Arduino Soil Moisture Detector

In this hands-on example project, you will learn how to build and operate a soil moisture detector. The soil moisture detector will use an RGB LED to indicate the moisture status of the soil. A green light will indicate that the moisture is ideal for plants, a blue light will indicate that there is too much water in the soil, and a red light will indicate that the soil is too dry.

Parts List

To build this example project, you will need

- 1 water detector
- 1 RGB LED
- 1 breadboard (optional to hold water detector)
- Wires to connect the components to the Arduino

Setting Up the Hardware

To build the hardware for this hands-on example project, you will need to:

- Connect the VCC pin on the water detector to the 5-V pin of the Arduino.
- Connect the GND pin on the water detector to the GND pin of the Arduino.
- Connect the AO pin on the water detector to analog pin A0 of the Arduino.

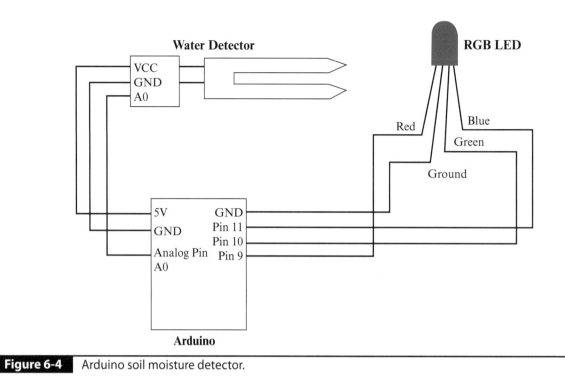

Figure 6-4 Arduino soil moisture detector.

- Connect the red terminal of the RGB LED to digital pin 9 of the Arduino.

- Connect the ground terminal of the RGB LED to a GND pin of the Arduino.

- Connect the green terminal of the RGB LED to digital pin 10 of the Arduino.

- Connect the blue terminal of the RGB LED to digital pin 11 of the Arduino (Figure 6-4).

Setting Up the Software

This hands-on example project builds on the preceding hands-on example project involving the water detector alarm. The additions and changes to the preceding code listing are shown in bold type. In addition, some code was eliminated, such as the code dealing with the buzzer and the push button.

The additions and modifications of the code include:

1. The blue terminal of the RGB LED is assigned to pin 11 of the Arduino.

2. The green terminal of the RGB LED is assigned to pin 10 of the Arduino.

3. The red terminal of the RGB LED is assigned to pin 9 of the Arduino.

4. In the `setup()` function, a new text message is printed to the Serial Monitor indicating that the soil moisture program has started.

5. The `Reset()` function is defined, and it:

   a. Sets the red color component of the RGB LED to 0.

   b. Sets the green color component of the RGB LED to 0.

   c. Sets the blue color component of the RGB LED to 0.

6. The `SetRGBLED()` function is defined, and it does the following:

   a. If the soil is wet or very wet, it sets the blue component of the RGB LED to 50.

   b. If the soil is dry or very dry, it sets the red component of the RGB LED to 50.

c. If the soil is damp, it sets the green component of the RGB LED to 50. For this example project, the desired target range of moisture is between WET and DRY.

7. In the loop() function, the following additions were made:

a. The red, green, and blue color components of the RGB LED are set to 0 by calling the Reset() function.

b. The red, green, and blue color components of the RGB LED are set based on the value read from the water detector by calling the SetRGBLED(RawValue) function. The input parameter is the RawValue variable that was read from the water detector. See Listing 6-4.

Listing 6-4 Arduino Soil Moisture Detector

```
// Soid Moisture Detector
const int VERY_WET = 400;
const int WET = 500;
const int DAMP = 700;
const int DRY = 850;
const int VERY_DRY = 950;

int DetectorPin = A0;
int RawValue = 0;

// RGB value limits are 0-255
int BluePin = 11;
int GreenPin = 10;
int RedPin = 9;

void setup()
{
 pinMode(DetectorPin, INPUT);
 Serial.begin(9600);
 Serial.println("Soil Moisture Detector
   ...");
}

void PrintMoistureStatus(int value)
{
```

```
Serial.print("MoistureStatus: ");
if (value <= VERY_WET)
{
    Serial.print("VERY_WET");
}
else
if (value <= WET)
{
    Serial.print("WET");
}
else
if (value <= DAMP)
{
    Serial.print("DAMP");
}
else
if (value <= DRY)
{
    Serial.print("DRY");
}
else
{
    Serial.print("VERY_DRY");
}
}

void Reset()
{
 // Clear All Colors
 analogWrite(RedPin, 0);
 analogWrite(GreenPin, 0);
 analogWrite(BluePin, 0);
}

void SetRGBLED(int value)
{
 if (value <= WET)
 {
    analogWrite(BluePin, 50);
 }
 else
 if (value >= DRY)
 {
    analogWrite(RedPin, 50);
 }
 else
 {
    analogWrite(GreenPin,50);
```

(continued on next page)

Listing 6-4   Arduino Soil Moisture Detector
(*continued*)

```
  }
}

void loop()
{
 RawValue = analogRead(DetectorPin);
 PrintMoistureStatus(RawValue);
 Serial.print(" , RawValue: ");
 Serial.println(RawValue);

 Reset();
 SetRGBLED(RawValue);
}
```

Running the Program

Upload the program to your Arduino, and start the Serial Monitor. The RGB LED should show red to indicate that the soil is very dry because there should be nothing between the soil probes. Wrap a wet towel around the soil probe portion of the sensor. Press down on the towel, and you should see the RGB LED change from red to green, which means that the soil is damp and has the correct amount of moisture. Press down harder, and you should see the LED change to blue, which means that the soil is too wet. You may need to experimentally determine what values you will need for these various categories by actually putting the soil moisture probe into the soil, pouring an adequate amount of water into the soil, and reading the corresponding value from the Serial Monitor and then changing the values in the program. You will then need to upload the program to the Arduino.

Light Detector (Photo Resistor)

A light detector or photo resistor is basically a variable resistor that can change its level

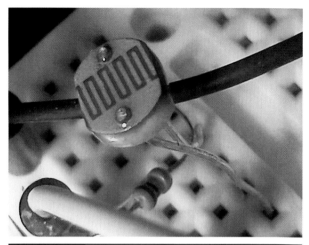

Figure 6-5   The photo resistor.

of resistance based on the amount of light it detects. Recall that voltage across a resistor is

$$V = IR \quad \text{or} \quad \text{Voltage} = \text{current} \times \text{resistance}$$

Thus, when the resistance of the light detector changes, the voltage across the light detector also changes. The light detector or photo resistor can be easily distinguished by the square-wave type of pattern on top of the component (Figure 6-5).

Hands-on Example: Arduino Rooster Alarm

In this hands-on example project, you will learn how to build a rooster alarm that produces sounds using a piezo buzzer when the light level in the environment has risen to a certain point. The alarm can be reset by pressing a button. The current level of lighting is displayed in real time using the Serial Monitor.

Parts List

To build this hands-on example project, you will need

- 1 light detector (photo resistor)
- 2 10-kΩ resistors
- 1 piezo buzzer

- 1 push button
- 1 breadboard (to hold the light detector and push button)

Setting Up the Hardware

To build this example project, you will need to:

- Connect one end of the light detector (photo resistor) to the 5-V pin of the Arduino.

- Connect the other end of the light detector to a node that contains a 10-kΩ resistor and a wire that goes to analog pin A0 of the Arduino. Connect the other end of the resistor to a GND pin of the Arduino.

- Connect the negative terminal on the buzzer to a GND pin of the Arduino.

- Connect the positive terminal on the buzzer to digital pin 9 of the Arduino.

- Connect one terminal on the push button to the 3.3-V pin of the Arduino.

- Connect the other terminal on the push button to a node that contains a 10-kΩ resistor and a wire to digital pin 7 of the Arduino. Connect the other end of the 10-kΩ resistor to a GND pin of the Arduino (Figure 6-6).

Setting Up the Software

To determine the level of lighting, measure the voltage across a 10-kΩ resistor that is attached in series with the light detector. The light detector is attached to the 5-V output from the Arduino, and the resistor is attached to GND on the Arduino (Figure 6-6). As the light level changes, the level of resistance of the light detector changes, and thus the voltage across the light detector changes. Because the total voltage must be 5 V for the two components, the voltage across the 10-kΩ resistor also will change. The analog voltage readings will be mapped into values from 0 to 1,023.

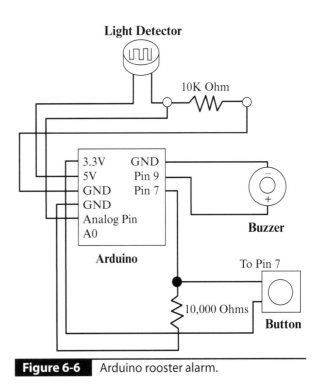

Figure 6-6   Arduino rooster alarm.

I experimentally determined some analog voltage readings that correspond to various lighting levels. I determined that direct sunlight is in the range of 900 to 1,023:

```
const int DirectSunlightCutOff = 1023;
```

Bright light was determined to be from 525 to 900:

```
const int BrightLightCutOff = 900;
```

Medium light was determined to be from 310 to 525:

```
const int MediumLightCutOff = 525;
```

Low lighting was determined to be between 200 and 310:

```
const int LowLightCutOff = 310;
```

Very low lighting was determined to be between 100 and 200:

```
const int VeryLowLightCutOff = 200;
```

Nighttime lighting was determined to be between 0 and 100:

```
const int NightCutOff = 100;
```

The node that consists of the light detector and the 10-kΩ resistor and measures the light levels in terms of voltage is assigned to analog pin A0 of the Arduino:

```
int DetectorPin = A0;
```

The `RawValue` variable holds the data read in from the light detector and is initialized to 0:

```
int RawValue = 0;
```

The buzzer is assigned to digital pin 9 of the Arduino:

```
int BuzzerPin = 9;
```

The frequency of the sound generated by the buzzer is set to 300 hertz:

```
int BuzzerFreq = 300;
```

The buzzer pitch range for the rooster call alarm is 400 hertz:

```
int BuzzerFreqRange = 400;
```

The number of "rooster calls" to sound when the sun comes up is set to 5:

```
int NumberRoosterCalls = 5;
```

If the `RoosterCallsFinished` variable is true, then the alarm, or "rooster," has finished its five "rooster calls," and the alarm stops. It is initialized to false to indicate that the alarm has not been tripped yet by the rising of the sun:

```
int RoosterCallsFinished = false;
```

The push button is assigned to digital pin 7 of the Arduino:

```
int ButtonPin = 7;
```

The `AlarmTripped` variable is true if the sun has risen and the alarm has been tripped; otherwise, it is false:

```
boolean AlarmTripped = false;
```

The `setup()` function initializes the Serial Monitor with a communication speed of 9,600 baud and prints out a text message to the Serial Monitor indicating that the program has started. See Listing 6-5.

Listing 6-5 The `setup()` Function

```
void setup()
{
 Serial.begin(9600);
 Serial.println("Rooster Alarm ...");
}
```

The `PrintBrightnessLevel()` function prints a text message to the Serial Monitor indicating the light level based on the value read from the light detector. The value read from the light detector is compared with a series of "cutoff" values that indicate the end of the range for that lighting level. For example, the `NightCutOff` value is the highest value for the night light level. See Listing 6-6.

Listing 6-6 The `PrintBrightnessLevel()` Function

```
void PrintBrightnessLevel(int value)
{
 Serial.print("BrightnessLevel: ");
 if (value <= NightCutOff)
 {
    Serial.print("Night");
 }
 else
 if (value <= VeryLowLightCutOff)
 {
    Serial.print("Very Low Light");
 }
 else
 if (value <= LowLightCutOff)
 {
    Serial.print("Low Light");
 }
 else
 if (value <= MediumLightCutOff)
 {
    Serial.print("Medium Light");
 }
 else
 if (value <= BrightLightCutOff)
 {
    Serial.print("Bright Light");
```

```
}
else
{
    Serial.print("Direct Sun Light");
}
}
```

The `RoosterCall()` function sounds the number of "rooster calls" in the `NumberRoosterCalls` variable. A "rooster call" is meant to simulate the actual sound a rooster makes when the sun rises by increasing the pitch of the sound using the buzzer. After all the "rooster calls" are completed, the buzzer sounds are terminated. See Listing 6-7.

Listing 6-7 The `RoosterCall()` Function

```
void RoosterCall()
{
  for (int i = 0; i < NumberRoosterCalls;
    i++)
  {
    for (int j = 0; j < BuzzerFreqRange;
      j++)
    {
      tone(BuzzerPin, BuzzerFreq + j);
      delay(10);
    }
  }
noTone(BuzzerPin);
RoosterCallsFinished = true;
}
```

The `ResetSoundConfirmation()` function causes the buzzer to emit a series of sounds that indicate that the alarm reset button has been pressed by doing the following five times:

1. Emitting a tone using the buzzer.

2. Holding this tone for 100 milliseconds.

3. Stopping the tone from being emitted by the buzzer.

4. Suspending program execution for 100 milliseconds. See Listing 6-8.

Listing 6-8 The `ResetSoundConfirmation()` Function

```
void ResetSoundConfirmation()
{
  for (int i = 0; i < 5; i++)
  {
    tone(BuzzerPin, BuzzerFreq);
    delay(100);

    noTone(BuzzerPin);
    delay(100);
  }
}
```

The `loop()` function holds the main rooster alarm logic and continuously monitors the light level by:

1. Reading the status of the reset button.

2. If the button is being pressed, then the function causes the buzzer to emit the reset sound sequence and resets the alarm so that it can be triggered again by the light level in the environment.

3. Reading in the light level from the light detector.

4. Printing out the light level description to the Serial Monitor by calling the `PrintBrightnessLevel()` function.

5. Printing out the raw numerical value of the light level to the Serial Monitor.

6. If the light level is greater than or equal to low light, which indicates that the sun is rising, then the function trips the alarm. This means that the light level is greater than or equal to `VeryLowLightCutOff`, and thus the light level is low or greater.

7. If the alarm has been tripped and the "rooster call" sound effects have not been completed, then the function starts the "rooster call" sound effects by executing the `RoosterCall()` function. See Listing 6-9.

Listing 6-9 The `loop()` Function

```
void loop()
{
 // Read Reset Button
 RawValue = digitalRead(ButtonPin);
 if (RawValue == 1)
 {
    ResetSoundConfirmation();
    AlarmTripped = false;
    RoosterCallsFinished = false;
 }

 RawValue = analogRead(DetectorPin);
 PrintBrightnessLevel(RawValue);
 Serial.print(" , RawValue: ");
 Serial.println(RawValue);

 if (RawValue >= VeryLowLightCutOff)
 {
    AlarmTripped = true;
 }

 if (AlarmTripped && !RoosterCallsFinished)
 {
    RoosterCall();
 }
}
```

Running the Program

First, dim the lights in the room you are in so that they are as low as possible. Upload the program to your Arduino, and start the Serial Monitor. Ideally, the light is low enough that the alarm does not activate. If you are in a room with variable lighting, adjust the brightness of the lights in increments until the alarm triggers. You should hear five "rooster calls" or sounds from the buzzer where the tone increases from the beginning pitch. After the five "rooster calls" are completed, the sounds should stop. Turn off the lights, and press the push button. This should reset the alarm so that it can be triggered by light again.

Sound Detector

A sound detector senses noise in the environment. The type of sound detector we will use in this book is a digital sound detector that is activated when a certain level of sound in the environment is sensed. A small microphone is attached to the sensor board to pick up sounds (Figure 6-7).

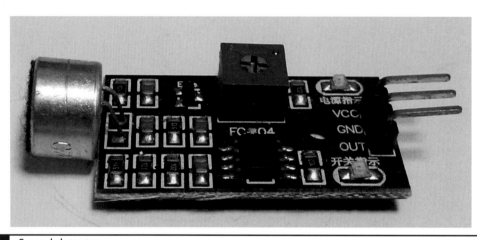

Figure 6-7 Sound detector.

Hands-on Example: Arduino Light Clapper

In this hands-on example project, you will learn how to build a system using a sound detector that enables you to turn a led on and off by clapping two times in quick succession. The operation is similar to a commercially available product known as the Clapper, which is a sound-activated electrical switch that can activate or deactivate lights that are plugged into it when the user claps.

Parts List

The parts you will need to build this example project include

- 1 sound detector sensor
- 1 LED
- 1 breadboard (optional to hold sound detector)
- Wires to connect the components together

Setting Up the Hardware

To build this hands-on example project, you will need to:

- Connect the VCC pin on the sound detector to the 3.3-V pin of the Arduino.
- Connect the GND pin on the sound detector to a GND pin of the Arduino.
- Connect the output pin on the sound detector to digital pin 8 of the Arduino.
- Connect the negative terminal of the LED to a GND pin of the Arduino.
- Connect the positive terminal of the LED to digital pin 7 of the Arduino (Figure 6-8).

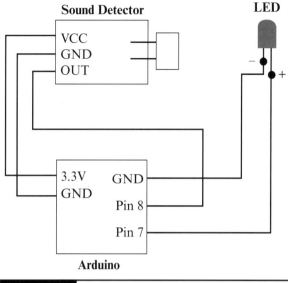

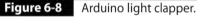

Figure 6-8 Arduino light clapper.

Setting Up the Software

The `NumberSounds` variable holds the total number of sounds detected by the sound detector since the program started and is initialized to 0:

```
int NumberSounds = 0;
```

The output pin on the sound detector is assigned to digital pin 8 of the Arduino:

```
int SoundPin = 8;
```

The `RawValue` variable holds the status of the sound detector and is initialized to 0:

```
int RawValue = 0;
```

The `NumberClaps` variable holds the number of claps that are currently being counted toward the two claps needed to toggle the LED light. This is initialized to 0 claps:

```
int NumberClaps = 0;
```

The `LightOn` variable is nonzero if the LED is turned on and 0 otherwise. Here it is initialized to 0:

```
int LightOn = 0;
```

The `SoundDetectedTime` variable holds the time of the last sound detected and is used to determine the number of unique claps because a single clap can generate more than one positive sound detector reading. This variable is initialized to 0:

```
unsigned long SoundDetectedTime = 0;
```

The `PreviousSoundDetectedTime` variable holds the time of the previous sound detected and is used to determine the number of unique claps:

```
unsigned long PreviousSoundDetectedTime
= 0;
```

The `UniqueClapMinTime` variable holds the minimum time in milliseconds between detected sounds that is required for the program to recognize a sound detected by the sound detector as a unique clap:

```
int UniqueClapMinTime = 100;
```

The `LEDPin` variable represents digital pin 7, which is connected to the positive terminal of the LED:

```
int LEDPin = 7;
```

The `PreviousClapTime` variable holds the time in milliseconds that the previous clap occurred and is initialized to 0:

```
unsigned long PreviousClapTime = 0;
```

The `CurrentClapTime` variable holds the time in milliseconds that the current clap has occurred and is initialized to 0:

```
unsigned long CurrentClapTime = 0;
```

The `MaxTimeBetweenClaps` variable holds the maximum time between two consecutive claps that can indicate a command to toggle the LED light. The default is set to 2,000 milliseconds, or 2 seconds.

```
unsigned long MaxTimeBetweenClaps = 2000;
```

The `setup()` function initializes the program and:

1. Sets the Arduino pin that is connected to the sound detector's output pin to be an input pin so that voltages can be measured.

2. Sets the Arduino pin that is connected to the LED's positive terminal to be an output

pin that it can deliver voltage to and drive the LED.

3. Initializes the Serial Monitor and sets the communication speed to 9,600 baud.

4. Prints a text message to the Serial Monitor indicating that the program has started. See Listing 6-10.

Listing 6-10   The `setup()` Function

```
void setup()
{
 pinMode(SoundPin, INPUT);
 pinMode(LEDPin, OUTPUT);
 Serial.begin(9600);
 Serial.println("The Light Clapper ...");
}
```

The `IsSoundPartOfUniqueClap()` function determines whether the sound detected by the sound detector is the start of a new unique clap or is part of the current clap. The function does this by:

1. Calculating the elapsed time since the previous sound was detected.

2. If this time is greater than or equal to the minimum time required for a unique clap to be recognized, then the function returns 1.

3. Otherwise, the function returns 0. See Listing 6-11.

Listing 6-11   The `IsSoundPartOfUniqueClap()` Function

```
int IsSoundPartOfUniqueClap()
{
 int result = 0;
 unsigned long ElapsedTime =
   SoundDetectedTime -
   PreviousSoundDetectedTime;
 if (ElapsedTime >= UniqueClapMinTime)
 {
   result = 1;
 }
 return result;
}
```

The `CheckTurnOnOffLight()` function determines whether the LED needs to be toggled by determining if there have been two consecutive claps detected within 2 seconds. The function does this by:

1. Calculating the elapsed time between the current detected clap and the previous detected clap.

2. If the elapsed time is less than or equal to the maximum allowed time between claps, then if the number of claps detected is two, sets the return value to 1, which means that the LED on/off status should be toggled, and resets the number of claps detected to 0.

3. If the elapsed time is greater than the maximum allowed time between claps, sets the number of claps detected to one to indicate that the currently detected clap is the only valid clap for the two-clap turn on/off sequence.

4. Returns 1 if the LED should be toggled and 0 otherwise. See Listing 6-12.

Listing 6-12 The `CheckTurnOnOffLight()` Function

```
int CheckTurnOnOffLight()
{
 int result = 0;
 unsigned long ElapsedTime =
   CurrentClapTime - PreviousClapTime;
 if (ElapsedTime <= MaxTimeBetweenClaps)
 {
    if (NumberClaps == 2)
    {
       result = 1;
       NumberClaps = 0;
    }
 }
 else
 {
    NumberClaps = 1;
 }
 return result;
}
```

The `loop()` function reads the output from the sound detector, determines whether the sound detected is a unique clap or part of a previous clap, determines if a two-clap sequence has been detected, and if so, toggles the LED light. The function does this by:

1. Reading the status of the sound detector.

2. If the value read is equal to 0, which means that a sound has been detected, then the function:

   a. Prints out some debug information to the Serial Monitor.

   b. Updates the variables that keep track of the times for the current sound detection event and the previous sound detection event.

3. If this sound detection event is part of a new unique clap, then the function:

   a. Increases the number of claps that have been detected by 1.

   b. Updates the times for the previous clap and the current clap.

4. If two consecutive claps have occurred within 2 seconds, then the function:

   a. Toggles the LED status variable by performing a bitwise NOT operation on the `LightOn` variable.

   b. Turns on the LED if the `LightOn` variable evaluates to true (nonzero).

   c. Turns off the LED if the `LightOn` variable evaluates to false (zero). See Listing 6-13.

Listing 6-13 The `loop()` Function

```
void loop()
{
 RawValue = digitalRead(SoundPin);
 if (RawValue == 0)
 {
    Serial.print("SOUND DETECTED ... ");
    Serial.print(", Sound Number: ");
    Serial.print(NumberSounds);
    Serial.print(", RawValue: ");
    Serial.println(RawValue);
    NumberSounds++;

    // Process raw data for claps
    PreviousSoundDetectedTime =
      SoundDetectedTime;
    SoundDetectedTime = millis();
    if(IsSoundPartOfUniqueClap())
    {
       NumberClaps++;

       // Update Clap Times
       PreviousClapTime =
         CurrentClapTime;
       CurrentClapTime = millis();

       // Turn Light ON/OFF as needed
       if (CheckTurnOnOffLight())
       {
          LightOn = ~LightOn;
          if (LightOn)
          {
             digitalWrite(LEDPin, HIGH);
          }
          else
          {
             digitalWrite(LEDPin, LOW);
          }
       }
    }
 }
}
```

Running the Program

Upload the program to your Arduino, and start the Serial Monitor. Clap once. You should see output similar to:

```
The Light Clapper ...
SOUND DETECTED ... , Sound Number: 0,
   RawValue: 0
SOUND DETECTED ... , Sound Number: 1,
   RawValue: 0
```

The LED should still be off because you have not clapped twice. Clap twice quickly, and the LED should turn on. Then clap twice quickly, and the LED should turn off. Clap once but wait 4 seconds and clap again. The LED should remain off because the time interval between claps has been exceeded.

Hands-on Example: Raspberry Pi "Out of Breath" Game

In this hands-on example project, you will learn how to make a sound game called "Out of Breath." The objective of this game is to create a steady sound with your voice that will be detected by a sound detector. When the sound detector detects a sound, it will increment a counter. When the sound detector does not detect a sound, it will decrement this counter. The goal is to generate a sound long enough to reach a target number that is entered by the user.

Parts List

To construct this hands-on example project, you will need

- 1 sound detector

- 1 LED

- 1 breadboard (optional to hold the sound detector)

■ Wires to connect the components to the Raspberry Pi

Setting Up the Hardware

To build the hardware for this example project, you will need to:

■ Connect the VCC pin on the sound detector to the 3.3-V pin of the Raspberry Pi.

■ Connect the GND pin on the sound detector to a GND pin of the Raspberry Pi.

■ Connect the output pin on the sound detector to pin 15 of the Raspberry Pi.

■ Connect the negative terminal of the LED to a GND pin of the Raspberry Pi.

■ Connect the positive terminal of the LED to pin 14 of the Raspberry Pi (Figure 6-9).

Setting Up the Software

This example project asks you to enter a number and then produce a sound using your voice that is picked up by the sound detector. The program increments a number every time it detects a sound and decrements the number every time it does not detect a sound. The objective is to produce the sound until the game counter

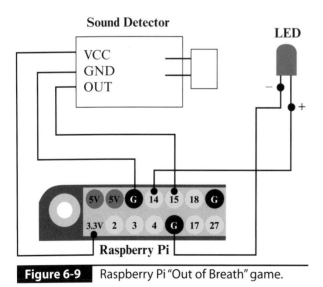

Figure 6-9   Raspberry Pi "Out of Breath" game.

matches the number you entered. The game then ends, displays the total time, and starts over. The program does this by:

1. Importing the RPi.GPIO library, which contains the functions that control the GPIO pins on the Raspberry Pi.

2. Importing the time library, which contains the functions that relate to measuring the passage of time.

3. Importing the pygame library, which contains sound effects–related code.

4. Assigning the output pin on the sound detector to pin 15 of the Raspberry Pi.

5. Assigning the positive terminal of the LED (`LEDPin`) to pin 14 of the Raspberry Pi.

6. Assigning to the variable that controls the frequency of the LED's pulse-width modulation (`Freq`) a value of 50.

7. Setting the variable (`GameIncrement`) that controls the amount to increment the number entered by the user when sound is detected to 0.1.

8. Setting the variable (`GameDecrement`) that controls the amount to decrement the number entered by the user when sound is not detected to 0.01.

9. Setting the Raspberry Pi's pin numbering method to BCM mode.

10. Setting the Raspberry Pi pin that is connected to the positive terminal of the LED to be an output pin so that voltage and current can be delivered to the LED.

11. Setting the Raspberry Pi pin that is connected to the output pin on the sound detector to be an input pin where voltages can be measured and thus data from the sound detector can be read.

12. Initializing the pygame system.

13. Initializing the pygame sound system.

14. Creating a sound object called `Alarm` from a .wav file located in the same directory from which the program is run.

15. Initializing pulse-width modulation (PWM) for the Raspberry Pi pin that is connected to the positive terminal of the LED.

16. Starting the PWM on the pin and setting the value to 0 brightness.

17. Doing the following until the user presses the CTRL-C keys to stop program execution:

    a. Requesting a target number from the user.

    b. Getting the current time and saving it in the `StartTime` variable.

    c. Setting the variable `GameNumber` to 0.

    d. Defining the variable that converts the `GameNumber` to a level of brightness on the LED. A value of 100 for brightness is equivalent to the target number entered by the user in step 1.

    e. Setting the brightness of the LED to 0, which is off.

18. As long as the `GameNumber` is less than the `TargetNumber` entered by the user:

    a. Reading in the status of the sound detector.

    b. Increasing the value of `GameNumber` if sound was detected.

    c. Decreasing the value of `GameNumber` if sound was not detected.

    d. Setting the `GameNumber` variable to 0 if it is less than 0.

    e. Calculating the brightness of the LED by multiplying the `GameNumber` variable by the `Conversion` variable.

    f. Restricting the range of brightness for the LED to the range of 0 to 100.

    g. Changing the brightness of the LED.

    h. Printing out the current `GameNumber` to the Serial Monitor.

19. Playing the sound represented by the `Alarm` object, which indicates that the player has succeeded in reaching the target number.

20. Getting the current time, which represents the time that the game has ended.

21. Printing out a text message to the Serial Monitor that indicates that the player has won the game and printing out the elapsed time.

22. Printing out a text message to the Serial Monitor that indicates that the program is exiting.

23. Stopping the PWM on the LED pin.

24. Deallocating resources for the GPIO pins. See Listing 6-14.

Listing 6-14 "Out of Breath" Game

```
# PI Out of Breath Game
import RPi.GPIO as GPIO
import time
import pygame


DetectorPin = 15
LEDPin = 14
Freq = 50


GameIncrement = 0.1
```

```
GameDecrement = 0.01

# Setup
GPIO.setmode(GPIO.BCM)
GPIO.setup(LEDPin, GPIO.OUT)
GPIO.setup(DetectorPin, GPIO.IN)
#SFX
pygame.init()
pygame.mixer.init()
Alarm = pygame.mixer.Sound("match1.wav")

PWM = GPIO.PWM(LEDPin, Freq)
PWM.start(0)

try:
 while 1:
    TargetNumber = int(input("Please Enter Target Number: "))
    StartTime = time.clock()
    GameNumber = 0
    # Convert Number into LED brightness
    Conversion = 100.0/TargetNumber
    # Reset Brightness of LED to off
    PWM.ChangeDutyCycle(0)
    while (GameNumber < TargetNumber):
       RawValue = GPIO.input(DetectorPin)
       if (RawValue == 0):
          # Sound Detected
          GameNumber = GameNumber + GameIncrement
       else:
          # No Sound
          GameNumber = GameNumber - GameDecrement
       if (GameNumber < 0):
          GameNumber = 0
       Brightness = GameNumber * Conversion
       # Range Check Brightness of LED
       if (Brightness < 0):
          Brightness = 0
       if (Brightness > 100):
          Brightness = 100
       PWM.ChangeDutyCycle(Brightness)
       print("GameNumber: ", GameNumber)
    Alarm.play()
    EndTime = time.clock()
    print("You Did It! Total Time:" , EndTime - StartTime);
except KeyboardInterrupt:
 pass
print ("...... Exiting Program ......")
PWM.stop()
GPIO.cleanup()
```

Running the Program

Run the program using Python from the Raspberry Pi's terminal program. Enter a number such as 200, and press the RETURN key. Produce a sound, for example, by saying "Ahhhh" near the microphone on the sound detector. You should see an increase in the GameNumber variable displayed on the terminal as well as an increase in brightness of the LED. Stop making the sound for a moment, and notice the decrease in brightness of the LED and the decrease in the GameNumber variable. Start making sounds again until you have won the game. You should hear a sound effect, and there should be another prompt asking you for a target number. Try changing the value of the GameDecrement variable to make the game more challenging or easier.

Summary

This chapter introduced you to various environmental sensors for the Arduino and Raspberry Pi. First up was the water detector sensor. The hands-on example showed you how this sensor can be used to detect a water heater leak and to measure the moisture level in soil. Next, you were introduced to the light detector, also known as a photo resistor. This sensor was used in a hands-on example to simulate the call a rooster makes when the sun rises in the morning. Next came the sound detector sensor. This sensor was used in an Arduino program that "listens for" two claps in quick succession and toggles the state of a LED if the claps occur. Finally, you used a sound sensor with a Raspberry Pi to create a game that uses sound as a key element.

## CHAPTER 7

Human Sensor Projects

THIS CHAPTER COVERS SENSORS that can be used to detect humans directly or indirectly. In the first example, you will build an Arduino glass break alarm system using a sound detector. This is followed by a Raspberry Pi version of the system. Next is a discussion of the HC-SR501 infrared motion sensor, which uses changes in infrared radiation (heat) to detect motion. Then you will build an infrared motion detector alarm system for the Arduino, followed by a similar system for the Raspberry Pi. Next is a discussion of the HC-SR04 distance sensor, which measures distances up to 400 cm in length. Then you will build an intruder alarm system using this distance sensor with the Arduino. Finally, you will build an Arduino collision alarm system that also uses the distance sensor.

Hands-on Example: Arduino Glass Break Alarm

In this hands-on example, you will build a glass break alarm that activates a buzzer alarm whenever a sound is detected. The alarm will continue to sound until the Reset button is pressed. Pressing the Reset button will reset the alarm to its original state, where sounds will be detected again. This alarm system will detect not only glass breaking but other sounds, such as doors being kicked, voices, or other sounds in a room that will indicate that someone is present.

Parts List

To build this hands-on example, you will need

- 1 sound detector
- 1 piezo buzzer
- 1 push button
- 1 10,000-ohm resistor
- 1 breadboard (optional to hold the sound detector)

Setting Up the Hardware

To build the hardware for this example, you will need to:

- Connect the VCC pin on the sound detector to the 3.3-V pin on the Arduino.
- Connect the GND pin on the sound detector to a GND pin on the Arduino.
- Connect the OUT pin on the sound detector to pin 8 on the Arduino.
- Connect the negative terminal of the piezo buzzer to a GND pin on the Arduino.
- Connect the positive terminal of the piezo buzzer to pin 9 on the Arduino.
- Connect one pin on the push button to a node that includes a resistor and a wire that goes to pin 7 on the Arduino. Connect the other end of the resistor to a GND pin on the Arduino.
- Connect the other pin on the push button to the 3.3-V pin on the Arduino (Figure 7-1).

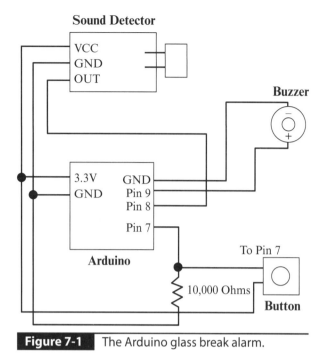

Figure 7-1   The Arduino glass break alarm.

Setting Up the Software

The glass break alarm detects an intruder by the noise he or she makes, including any noise from a broken window or a door that is kicked in or forced open. The alarm detects an intruder by:

1. Initializing to 0 the RawValue variable, which holds the data from the sound detector sensor.

2. Assigning to pin 8 on the Arduino the SoundPin variable, which represents the output pin on the sound detector.

3. Initializing to 0 the SoundCount variable, which holds the number of times the sound detector has detected a sound.

4. Initializing to 2 the Sensitivity variable, which holds the minimum number of times that a sound must be detected to trigger the alarm.

5. Assigning to pin 9 on the Arduino the BuzzerPin variable, which represents the positive terminal of the buzzer.

6. Initializing to 300 the BuzzerFreq variable, which represents the frequency of the tone emitted by the buzzer.

7. Assigning to pin 7 on the Arduino the ButtonPin variable, which is used to read the status of the button.

8. Calling the AlarmTripped variable. If the value is false, then the alarm has not been tripped. Otherwise, the value is true, which means that the alarm has been tripped and a sound has been detected.

9. Calling the setup() function, which initializes the program and:

    a. Sets the pin on the Arduino connected to the output pin on the sound detector to be an input pin so that data from the sound detector can be read in.

    b. Sets the pin on the Arduino connected to the push button to be an input pin so that the voltage across the 10-kΩ (kilo-ohm) resistor can be read in. The voltage is 1 if the push button has been pressed and 0 otherwise.

    c. Initializes the Serial Monitor with a communication speed of 9,600 baud.

    d. Prints a text message to the Serial Monitor indicating that the program has started.

10. Executing the loop() function repeatedly, which contains the main code and:

    a. Reads the status of the push button. If the push button has been pressed, the alarm is reset, and the number of sounds detected is set to 0.

    b. Determines the status of the sound detector. If the sound detector has detected a sound, then the number of sounds that have been counted is increased.

c. Then prints a text message to the Serial Monitor indicating that a sound has been detected and the number of sounds that have been detected.

d. If the number of sounds that have been counted is equal to or greater than the sensitivity value, then sets the alarm to be tripped.

e. If the alarm has been tripped, then produces a sound with the buzzer.

f. If the alarm has not been tripped, then stops any sound from the buzzer that is being produced.

See Listing 7-1.

Listing 7-1 The Glass Break Alarm

```
int RawValue = 0;

// Sound Detector
int SoundPin = 8;
int SoundCount = 0;
int Sensitivity = 2;

// Buzzer
int BuzzerPin = 9;
int BuzzerFreq = 300;

// Button
int ButtonPin = 7;

// Has alarm been triggered?
boolean AlarmTripped = false;

void setup()
{
 pinMode(SoundPin, INPUT);
 pinMode(ButtonPin, INPUT);

 Serial.begin(9600);
 Serial.println("Glass Break Alarm
   ...");
}

void loop()
```

```
{
 // Read Reset Button
 RawValue = digitalRead(ButtonPin);
 if (RawValue == 1)
 {
   // Reset alarm
   AlarmTripped = false;
   SoundCount = 0;
 }

 // Read Sound Detector
 RawValue = digitalRead(SoundPin);
 if (RawValue == 0)
 {
   Serial.print("SOUND DETECTED ...
     SoundCount: ");
   SoundCount++;
   Serial.println(SoundCount);
   if (SoundCount >= Sensitivity)
   {
     AlarmTripped = true;
   }
 }

 // Sound Alarm if alarm has been
 // tripped.
 if (AlarmTripped)
 {
   tone(BuzzerPin, BuzzerFreq);
 }
 else
 {
   noTone(BuzzerPin);
 }
}
```

Running the Program

Upload the program to the Arduino, and start the Serial Monitor. You should see a text message on the Serial Monitor indicating that the program has started:

```
Glass Break Alarm ...
```

Next, clap once, and you should see something like:

```
SOUND DETECTED ... SoundCount: 1
SOUND DETECTED ... SoundCount: 2
```

If the clap was detected by the sensor at least two times, then the buzzer alarm should activate. The alarm will continue until the push button is pressed, which resets the alarm to its original state.

Hands-on Example: Raspberry Pi Glass Break Alarm

In this hands-on example, you will build a glass break alarm for the Raspberry Pi using a sound detector sensor (which was discussed in Chapter 6). The sound detector will pick up the sound of breaking glass, a door that is forced open, or other loud sounds that indicate that a person is nearby. When the sound is detected, an audible alarm will activate until the Reset button is pressed. The Reset button restores the alarm system to its initial state.

Parts List

To complete this hands-on example project, you will need

- 1 sound detector (this was covered in Chapter 6)
- 1 push button
- 1 10-kΩ resistor
- 1 breadboard (optional for holding the sound detector)
- Wires to connect the components together

Setting Up the Hardware

To build this project, you will need to:

- Connect the VCC pin on the sound detector to a 3.3-Vt pin on the Raspberry Pi.
- Connect the GND pin on the sound detector to the GND pin on the Raspberry Pi.

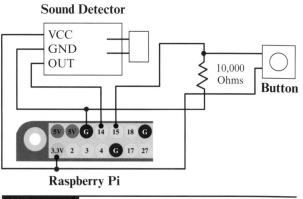

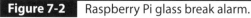

Figure 7-2   Raspberry Pi glass break alarm.

- Connect the OUT pin on the sound detector to pin 14 on the Raspberry Pi.
- Connect one pin on the push button to a node that contains a 10-kΩ resistor and a wire to pin 15 on the Raspberry Pi. Connect the other end of the 10-kΩ resistor to a GND pin on the Raspberry Pi.
- Connect the other terminal on the push button to a 3.3-V pin on the Raspberry Pi (Figure 7-2).

Setting Up the Software

The program for this example project implements the alarm by:

1. Importing the RPi.GPIO library, which is used to access the GPIO pins on the Raspberry Pi.

2. Importing the pygame library so that sounds can be played.

3. Initializing to 0 the RawValue variable, which holds the data that are read from the sound detector.

4. Initializing the SoundPin variable, which represents the output pin on the sound detector connected to pin 14 on the Raspberry Pi.

5. Initializing to 0 the SoundCount variable, which represents the number of sounds detected by the sound detector.

6. Initializing to 2 the `Sensitivity` variable, which represents the minimum number of detected sounds needed to trigger the alarm.

7. Initializing the `ButtonPin` variable, which is used to read the status of the push button connected to pin 15 on the Raspberry Pi.

8. Initializing to false or 0 the `AlarmTripped` variable, which is 1 if the alarm has been triggered by the detection of a sound.

9. Setting the pin numbering method for the Raspberry Pi to BCM mode.

10. Setting the pin on the Raspberry Pi that is connected to the output from the sound detector to be an input pin so that data can be read in.

11. Setting the pin on the Raspberry Pi that is used to read the status of the push button to be an input pin so that voltages can be measured.

12. Initializing the `pygame` module.

13. Initializing the `pygame` module's sound system.

14. Creating the sound object that represents the alarm sound effect from a file called "match1.wav" that is located in the same directory in which the Python script is being run.

15. Printing a text message to the Raspberry Pi's terminal screen indicating that the glass break alarm program is running.

16. Doing the following until the user creates a keyboard interrupt by pressing the CTRL-C combination:

    a. Reads the status of the Reset button. If the status is 1 (HIGH voltage), that is, the user is pressing down on the button, then reset the alarm system.

    b. Reads the status of the sound detector sensor. If a sound has been detected,

the status is 0 or LOW voltage. Then the program increases the count of the number of sounds that have been detected.

    c. Prints out to the Serial Monitor that a sound has been detected and the total number of sounds that have been detected so far.

    d. If the total number of sounds detected is equal to or greater than the sensitivity value, then the program trips the alarm.

    e. If the alarm has been tripped, then the program plays the alarm sound effect.

17. Prints out a text message to the Raspberry Pi terminal screen indicating that the program is terminating.

18. Releases the allocated resources related to the GPIO pins that were used in this program.

See Listing 7-2.

Listing 7-2 Raspberry Pi Glass Break Alarm

```
# Glass Break Alarm
import RPi.GPIO as GPIO
import pygame

RawValue = 0

# Sound Detector
SoundPin = 14
SoundCount = 0
Sensitivity = 2

# Button
ButtonPin = 15

# Has alarm been triggered?
AlarmTripped = 0

# Setup
GPIO.setmode(GPIO.BCM)
```

(continued on next page)

Listing 7-2   Raspberry Pi Glass Break Alarm (*continued*)

```
GPIO.setup(SoundPin, GPIO.IN)
GPIO.setup(ButtonPin, GPIO.IN)

#SFX
pygame.init()
pygame.mixer.init()
Alarm = pygame.mixer.Sound("match1.wav")
print("Glass Break Alarm ...")

try:
 while 1:
    # Read Reset Button
    RawValue = GPIO.input(ButtonPin)
    if (RawValue):
       # Reset alarm
       AlarmTripped = 0
       SoundCount = 0

    # Read Sound Detector
    RawValue = GPIO.input(SoundPin)
    if (RawValue == 0):
       SoundCount = SoundCount + 1
       print("SOUND DETECTED ...
         SoundCount: ", SoundCount)
       if (SoundCount >= Sensitivity):
          AlarmTripped = 1

    # Sound Alarm if alarm has been
    # tripped.
    if (AlarmTripped):
       Alarm.play()
except KeyboardInterrupt:
 pass
print ("...... Exiting Program ......")
GPIO.cleanup()
```

Running the Software

Before you run the program, you will need to copy a .wav file format sound effect to the same directory from which you plan on running the Python script. Change the line in the program that defines the exact sound filename to the name of your own sound file, such as

```
Alarm = pygame.mixer.Sound("YourCustom
SoundFilenameHere.wav")
```

Next, run the Python program on your Raspberry Pi using the terminal. Create a noise (such as clapping your hands together), and the alarm should be tripped. Once the alarm has been tripped, the sound effect you specified for the `Alarm` object should start playing continuously. To stop the sound effect and reset the alarm, press the button and hold it down until the alarm sound stops playing.

HC-SR501 Infrared Motion Detector

The HC-SR501 infrared motion detector detects the movement of a human being by sensing changes in the environment due to body heat. The front of the sensor is covered by a plastic dome (Figure 7-3).

The dome is removable, and underneath there should be labels that identify which pins are for voltage input, ground, and output (Figure 7-4).

Figure 7-3   Infrared motion detector with cover.

Figure 7-4 Infrared motion sensor without front cover showing the pin labels.

Hands-on Example: Arduino Infrared Motion Detector Alarm

In this hands-on example, you will build an infrared motion detection alarm. When the detector is first turned on, it goes into a "Waiting" state where it is not active. During this waiting state, the buzzer emits a constant series of beeps. After all the beeps have been played, the alarm switches into an "Active" state. In the active state, the alarm can detect motion. If motion is detected, then the buzzer alarm emits a constant tone, and the LED lights up, the alarm status changing into a "Tripped" state. If the user presses the button, then the alarm is reset by stopping sounds from the buzzer and then setting the alarm back into a "Waiting" state. From this waiting state, the alarm again transitions to an "Active" state where it detects motion.

Parts List

To build this example project, you will need

- 1 infrared motion detector
- 1 piezo buzzer
- 1 LED
- 1 push button
- 1 10-kΩ resistor
- 1 breadboard (to hold the LED and create nodes for the ground)
- Wires to connect the components together

Setting Up the Hardware

To build this example project, you will need to:

- Connect the VCC pin on the infrared sensor to the 5-V pin on the Arduino.
- Connect the GND pin on the infrared sensor to a GND pin on the Arduino.
- Connect the OUT pin on the infrared sensor to pin 8 on the Arduino.
- Connect the negative terminal on the buzzer to a GND pin on the Arduino.
- Connect the positive terminal on the buzzer to pin 9 on the Arduino.
- Connect the negative terminal on the LED to a GND pin on the Arduino or a GND node on the breadboard.

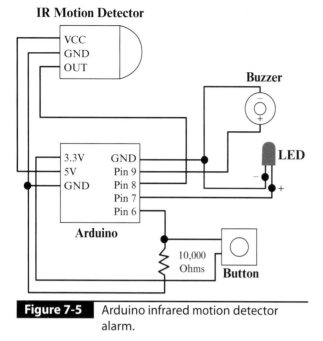

Figure 7-5 Arduino infrared motion detector alarm.

- Connect the positive terminal on the LED to pin 7 on the Arduino.

- Connect one terminal of the push button to a node that contains a 10-kΩ resistor and a wire that connects the node to pin 6 on the Arduino. Connect the other end of the resistor to the GND pin on the Arduino or a GND node on the breadboard.

- Connect the other terminal on the push button to the 3.3-V pin on the Arduino (Figure 7-5).

Setting Up the Software

The program for this example project is defined as follows:

1. The `SensorPin` variable represents the output pin on the infrared sensor and is assigned to pin 8 on the Arduino.

2. The `NumberDetections` variable represents the number of times the alarm has been tripped and is initialized to 0.

3. The `RawValue` variable holds the data read from the infrared sensor and is initialized to 0.

4. The `LEDPin` variable represents the positive terminal of the LED and is assigned to pin 7 on the Arduino.

5. The `BuzzerPin` variable represents the positive terminal on the piezo buzzer and is assigned to pin 9 on the Arduino.

6. The `BuzzerFreq` variable holds the frequency of the tone that will be generated by the buzzer when the alarm has been triggered and is initialized to 300.

7. The `BuzzerFreqWait` variable holds the frequency of the tone generated when the alarm is in the "Waiting" state and is initialized to 600.

8. The `ButtonPin` variable is used to read the status of the Reset button and is assigned to pin 6 on the Arduino.

9. The `AlarmState` enumeration defines the current status of the alarm system, which can be:

   a. **Waiting:** The alarm is not activated but is waiting for a certain time period. This time can be used by the user to exit a room before the alarm is activated.

   b. **Active:** The alarm is active and continuously monitoring the area for motion. In this state, the alarm can be tripped.

   c. **Tripped:** The alarm has been triggered by the motion of someone, and the buzzer should be emitting a sound and the LED should be lighted up. The alarm can be reset to the "Waiting" state by pressing the Reset button.

10. The `CurrentAlarmState` variable is an enumeration that holds the current status of the alarm system and is initialized to "Waiting."

11. The pin on the Arduino that is connected to the output pin on the infrared sensor is set

to be an input pin so that data can be read from the sensor.

12. The pin on the Arduino that is connected to the positive terminal of the LED is set as an output pin so that it can deliver voltage to the LED.

13. The pin that is used to read the status of the Reset button is set as an input pin so that voltages can be measured.

14. The Serial Monitor is initialized and set to a communication rate of 9,600 baud.

15. A text message is printed to the Serial Monitor indicating that the program has started.

16. The `ProcessWaitingState()` function is defined, and it:

    a. Generates 15 short beeps using the buzzer

    b. Sets the alarm state to "Active"

    c. Prints out a text message to the Serial Monitor indicating that the status of the alarm system is now active and ready to detect motion

17. The `loop()` function reads the status of the Reset button.

    a. If the Reset button has been pressed, the alarm status is changed to "Waiting," and a text message is printed to the Serial Monitor indicating that the status of the alarm has changed to "Waiting."

    b. If the current alarm state is "Waiting," then the program processes this waiting state by calling the `ProcessWaitingState()` function.

    c. If the current alarm state is "Active," then the program reads the status of the motion detector.

    d. If the motion sensor detects movement, then the program Increases the number of total motion detections by 1.

        i. Prints a text message to the Serial Monitor indicating that a motion has been detected and the total number of motions that have been detected since startup by the alarm system.

        ii. Sets the alarm system state to "Tripped."

        iii. Prints out a text message to the Serial Monitor indicating that the motion has been detected and the alarm has been tripped.

    e. If the current alarm system status is "Active," then the program turns on the LED. Otherwise, it turns off the LED.

    f. If the current alarm status is "Tripped," then the program generates a tone using the buzzer. Otherwise, it makes sure the buzzer is silent.

See Listing 7-3.

Listing 7-3 ■ Arduino Infrared Motion Detector Alarm

```
// IR Sensor
int SensorPin = 8;
int NumberDetections = 0;
int RawValue = 0;

// LED Indicator
int LEDPin = 7;

// Buzzer
int BuzzerPin = 9;
int BuzzerFreq = 300;
int BuzzerFreqWait = 600;

// Button
int ButtonPin = 6;
```

(continued on next page)

Listing 7-3 Arduino Infrared Motion Detector Alarm *(continued)*

```
enum AlarmState
{
 Waiting,
 Active,
 Tripped
};

AlarmState CurrentAlarmState = Waiting;

void setup()
{
 pinMode(SensorPin, INPUT);
 pinMode(LEDPin, OUTPUT);
 pinMode(ButtonPin, INPUT);

 Serial.begin(9600);
 Serial.println("IR Motion Sensor Alarm
   ...");
}

void ProcessWaitingState()
{
 for(int i = 0; i < 15; i++)
 {
    // Play Wait Tone
    tone(BuzzerPin, BuzzerFreqWait);
    delay (300);

    noTone(BuzzerPin);
    delay (300);
 }
 CurrentAlarmState = Active;
 Serial.println("CurrentAlarmState:
   Active");
}

void loop()
{
 // Read Reset Button
 RawValue = digitalRead(ButtonPin);
 if (RawValue)
 {
    CurrentAlarmState = Waiting;
    Serial.println("CurrentAlarmState:
      Waiting");
 }
```

```
 // Check if Alarm is in Waiting State
 if (CurrentAlarmState == Waiting)
 {
    ProcessWaitingState();
 }

 // Read Infrared Motion Sensor if
 // Alarm Active
 if (CurrentAlarmState == Active)
 {
    RawValue = digitalRead(SensorPin);
    if (RawValue == 1)
    {
      // Human Detected
      NumberDetections++;
      Serial.print("IR Motion Detector
        Triggered, NumberDetections: ");
      Serial.println(NumberDetections);

      CurrentAlarmState = Tripped;
      Serial.
        println("CurrentAlarmState:
        Tripped");
    }
 }

 // Set LED State
 if (CurrentAlarmState == Tripped)
 {
    digitalWrite(LEDPin, HIGH);
 }
 else
 {
    digitalWrite(LEDPin,LOW);
 }

 // Sound Alarm if alarm has been
 // tripped.
 if (CurrentAlarmState == Tripped)
 {
    tone(BuzzerPin, BuzzerFreq);
 }
 else
 {
    noTone(BuzzerPin);
 }
}
```

Running the Program

Upload the program to the Arduino, and open the Serial Monitor. The text indicating that the program has started will be displayed:

```
IR Motion Sensor Alarm ...
```

Next, the buzzer should emit a series of short beeps that indicate that the alarm is in the "Waiting" state, where the alarm will not be tripped by motion. After the series of beeps stops, the alarm will be put into "Active" mode, where the infrared motion sensor will detect movement. A text message stating that the alarm is now active should be displayed:

```
CurrentAlarmState: Active
```

Pass your hand in front of the motion sensor to activate the alarm. You should see the LED light up and the buzzer start to emit a constant tone. You should also see a text message on the Serial Monitor alerting you that the alarm has been tripped:

```
IR Motion Detector Triggered,
    NumberDetections: 1
CurrentAlarmState: Tripped
```

Now press the Reset button to return the alarm system to its original state. You should hear short beeps from the buzzer, and you should see a text message displayed on the Serial Monitor indicating that the alarm is in the "Waiting" state:

```
CurrentAlarmState: Waiting
```

Once all the beeps have been completed, the alarm system becomes active again and ready to detect more motion:

```
CurrentAlarmState: Active
```

Hands-on Example: Raspberry Pi Infrared Motion Detection Alarm

This example project shows you how to build an infrared motion detection alarm for the Raspberry Pi. The alarm will have three states: "Waiting," "Active," and "Tripped." In the "Waiting" state, the alarm does not detect motion. After 15 sound effects are played, the state of the alarm changes to "Active." In the "Active" state, the alarm does detect movement. In the "Tripped" state, a motion has been detected, and the alarm sound effect plays until the user hits the Reset button. The Reset button will restore the alarm system back to the "Waiting" state.

Parts List

For this example project, you will need

- 1 infrared motion detection sensor
- 1 LED
- 1 push button
- 1 10-kΩ resistor
- 1 breadboard
- Wires to connect the components

Setting Up the Hardware

To build this example project, you will need to:

- Connect the VCC pin on the motion detector to a 5-V pin on the Raspberry Pi.
- Connect the GND pin on the motion detector to a GND pin on the Raspberry Pi.
- Connect the OUT pin on the motion detector to pin 14 on the Raspberry Pi.
- Connect the negative terminal on the LED to a GND pin on the Raspberry Pi.
- Connect the positive terminal on the LED to pin 15 on the Raspberry Pi.
- Connect one terminal of the push button to a node that contains the 10-kΩ resistor and a wire to pin 18 on the Raspberry Pi. Connect the other end of the resistor to a GND pin on the Raspberry Pi.

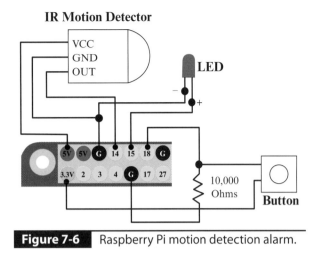

Figure 7-6 Raspberry Pi motion detection alarm.

■ Connect the other terminal on the push button to a 3.3-V pin on the Raspberry Pi (Figure 7-6).

Setting Up the Software

The program implements the alarm system by:

1. Importing the `RPi.GPIO` module into the program so that the functions related to the GPIO pins can be used.

2. Importing the `time` module into the program so that time delay functions can be accessed.

3. Importing the `pygame` module into the program so that sound effects can be used.

4. Assigning to pin 14 on the Raspberry Pi the `SensorPin` variable, which represents the output pin on the motion detector.

5. Initializing to 0 the `NumberDetections` variable, which holds the number of times the alarm has been tripped.

6. Initializing to 0 the `RawValue` variable, which holds the values read from the motion sensor.

7. Assigning to pin 15 on the Raspberry Pi the `LEDPin` variable, which represents the positive terminal on the LED.

8. Assigning to pin 18 on the Raspberry Pi the `ButtonPin` variable, which represents the alarm's Reset button.

9. Initializing the `CurrentAlarmState` variable, which holds the current state of the alarm, to "Waiting." In this state, the alarm does not detect motion.

10. Setting the GPIO pin numbering method on the Raspberry Pi to BCM mode.

11. Setting the pin on the Raspberry Pi that is connected to the positive terminal of the LED to be an output pin so that voltage can be placed on the LED.

12. Setting the pin on the Raspberry Pi that is connected to the output pin on the motion sensor to be an input pin so that data can be read from the motion sensor.

13. Setting the pin on the Raspberry Pi that is used to read the status of the Reset button to be an input pin.

14. Initializing the `pygame` system.

15. Initializing the `pygame` sound system.

16. Creating the `Alarm` sound object from a file that is played when the alarm system is tripped by motion.

17. Creating the `WaitSound` sound object from a file that is played when the alarm system is in the "Waiting" mode.

18. Printing out a text message to the terminal indicating that the program has started.

19. Defining the `ProcessWaitingState()` function, which produces 15 short sound effects using the `WaitSound` sound object that was created previously. The program switches the alarm system to "Active" mode and prints out a text message to the terminal indicating this change.

20. Doing the following until the user generates a keyboard interrupt by pressing the CTRL-C sequence:

a. Reads the status of the Reset button.

b. If the Reset button is pressed, then the program sets the current alarm state to "Waiting" and prints out a text message to the terminal indicating the change in state.

c. If the alarm status is "Waiting," then the program calls the `ProcessWaitingState()` function.

d. If the current alarm state is "Active," the program reads the current status of the motion detector.

e. If a motion is detected, the program:

   i. Increases the number of detected motions by 1.

   ii. Prints out the number of times the alarm has been tripped.

   iii. Sets the current state of the alarm to "Tripped."

   iv. Prints out the current state of the alarm.

f. If the alarm is tripped, the program turns on the LED; otherwise, it turns off the LED.

g. If the alarm is tripped, the program plays the alarm sound effect; otherwise, it stops the alarm sound effect if it is being played.

21. Deallocating resources related to the use of the GPiO pins on the Raspberry Pi before exiting.

See Listing 7-4.

Listing 7-4 Raspberry Pi Motion Detection Alarm

```
# Infrared Motion Sensor Alarm
import RPi.GPIO as GPIO
import time
import pygame

# IR Sensor
SensorPin = 14
NumberDetections = 0
RawValue = 0

# LED Indicator
LEDPin = 15

# Button
ButtonPin = 18
CurrentAlarmState = "Waiting"

# Setup
GPIO.setmode(GPIO.BCM)
GPIO.setup(LEDPin, GPIO.OUT)
GPIO.setup(SensorPin, GPIO.IN)
GPIO.setup(ButtonPin, GPIO.IN)

#SFX
pygame.init()
pygame.mixer.init()
```

(continued on next page)

Listing 7-4 Raspberry Pi Motion Detection Alarm (*continued*)

```python
Alarm = pygame.mixer.Sound("match1.wav")
WaitSound = pygame.mixer.Sound("match2.wav")
print("IR Motion Sensor Alarm ...")

def ProcessWaitingState():
 global WaitSound
 global CurrentAlarmState

 for counter in range(15):
    # Play Wait Tone
    WaitSound.play()
    time.sleep(0.5)
    WaitSound.stop()
    time.sleep(0.5)
 CurrentAlarmState = "Active"
 print("CurrentAlarmState: Active")

try:
 while 1:
    # Read Reset Button
    RawValue = GPIO.input(ButtonPin)
    if (RawValue):
       CurrentAlarmState = "Waiting"
       print("CurrentAlarmState: Waiting")

    # Check if Alarm is in Waiting State
    if (CurrentAlarmState == "Waiting"):
       ProcessWaitingState()

    # Read Infrared Motion Sensor if Alarm Active
    if (CurrentAlarmState == "Active"):
       RawValue = GPIO.input(SensorPin)
       if (RawValue == 1):
          # Human Detected
          NumberDetections = NumberDetections + 1
          print("IR Motion Detector Triggered, NumberDetections: ",NumberDetections)
          CurrentAlarmState = "Tripped"
          print("CurrentAlarmState: Tripped")

    # Set LED State
    if (CurrentAlarmState == "Tripped"):
       GPIO.output(LEDPin, GPIO.HIGH)
    else:
       GPIO.output(LEDPin, GPIO.LOW)

    # Sound Alarm if alarm has been tripped.
    if (CurrentAlarmState == "Tripped"):
       Alarm.play()
    else:
       Alarm.stop()
except KeyboardInterrupt:
 pass
GPIO.cleanup()
```

Running the Program

Before you run this program, you will need to install the two sound files that will represent the sounds for the alarm and for the beeps produced during the "Waiting" state. You will also need to change the filenames in the program code to match the files that you have copied to the same directory as the program file. Then run the program using Python from the Raspberry Pi's terminal screen. The text indicating that the program has started will be displayed:

```
IR Motion Sensor Alarm ...
```

The speaker should produce the sound you specified for the waiting state. After the series of sounds stops, the alarm will be put into "Active" mode, where the infrared motion sensor will detect movement. A text message stating that the alarm is now active should be displayed:

```
CurrentAlarmState: Active
```

Pass your hand in front of the motion sensor to activate the alarm. You should see the LED light up, and the sound effect for the alarm should start playing. You should also see text messages on the terminal alerting you that the alarm has been tripped:

```
('IR Motion Detector Triggered,
    NumberDetections:', 1)
CurrentAlarmState: Tripped
```

Now press the Reset button to return the alarm system to its original state. You should hear a series of sound effects, and you should see a text message displayed on the terminal indicating that the alarm is in the "Waiting" state:

```
CurrentAlarmState: Waiting
```

Once all the sound effects have been completed, the alarm system becomes active again and ready to detect more motion:

```
CurrentAlarmState: Active
```

HC-SR04 Distance Sensor

The HC-SR04 distance sensor can be used to measure the distance from the sensor to an object. The sensor uses a sound to determine the distance to an object by generating a sound pulse and detecting the reflected sound, or echo, from that sound pulse when it is reflected by the object. The time between emitting the sound pulse and receiving the echo is proportional to the distance to the object (Figure 7-7).

Figure 7-7   HC-SR04 distance sensor.

To find the distance to the nearest object in front of an HC-SR04 sensor, you have to:

- Set the Trigger pin on the sensor from 0 or LOW to 1 or HIGH for 10 microseconds and then set the pin value back to 0 or LOW (Figure 7-8). The distance sensor will then emit a series of sound pulses that will be reflected back to the sensor when an object has been hit by the sound waves.

- Read the pulse from the Echo pin on the sensor. The time of the HIGH portion of the pulse is proportional to the distance to the object that has been detected. The distance in inches to an object that is detected is the time in microseconds of the HIGH portion of the pulse divided by 148 (Figure 7-9).

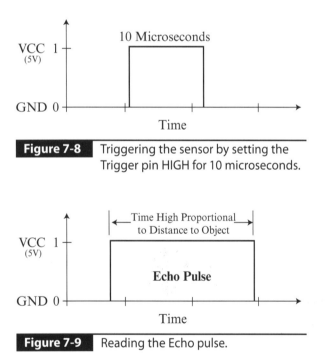

Figure 7-8   Triggering the sensor by setting the Trigger pin HIGH for 10 microseconds.

Figure 7-9   Reading the Echo pulse.

Hands-on Example: Arduino HC-SR04 Distance Sensor Intruder Alarm

For this example project, you will build an intruder alarm using an HC-SR04 distance sensor. The default distance is determined first. This is the distance from the sensor to any surrounding objects without an intruder present. Then you calculate a trip distance by subtracting an offset from the default distance. The trip distance represents the maximum distance to an object and will indicate that an intruder is present and cause the alarm to trip. The Reset button can be pressed to restore the alarm to its initial state, where the default distance and the trip distance will be determined first and then the alarm will enter an active state where intruders will be detected.

Parts List

To build this example project, you will need

- 1 HC-SR04 distance sensor
- 1 piezo buzzer
- 1 push button
- 1 10-kΩ resistor
- 1 breadboard (optional for distance sensor)
- Wires to connect the components to the Arduino

Setting Up the Hardware

To build this example project, you will need to:

- Connect the GND pin on the distance sensor to a GND pin on the Arduino.
- Connect the Trigger pin on the distance sensor to pin 8 on the Arduino.

- Connect the Echo pin on the distance sensor to pin 7 on the Arduino.

- Connect the VCC pin on the distance sensor to the 5-V pin on the Arduino.

- Connect the negative terminal on the buzzer to a GND pin on the Arduino.

- Connect the positive terminal on the buzzer to pin 9 on the Arduino.

- Connect one terminal of the push button to a node that contains a 10-kΩ resistor and a wire to pin 6 on the Arduino. Connect the other end of the resistor to a GND pin on the Arduino.

- Connect the other terminal on the push button to the 3.3-V pin on the Arduino (Figure 7-10).

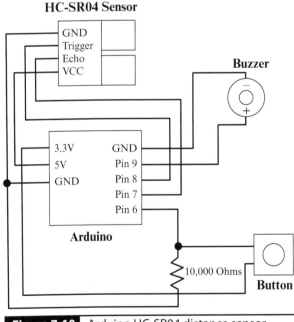

Figure 7-10 Arduino HC-SR04 distance sensor intruder alarm.

Setting Up the Software

The program that controls the alarm system in this example project:

1. Assigns to pin 8 on the Arduino the `TriggerPin` variable, which represents the Trigger pin on the distance sensor.

   ```
   int TriggerPin = 8;
   ```

2. Assigns to pin 7 on the Arduino the `EchoPin` variable, which represents the Echo pin on the distance sensor.

   ```
   int EchoPin = 7;
   ```

3. Initializes to 0 the `DurationPulse` variable, which holds the time in microseconds of the echo pulse generated by the sensor that will be used to determine the distance between the sensor and an object detected by the sensor.

   ```
   long DurationPulse = 0;
   ```

4. Initializes to 0 the `DistInches` variable, which holds the distance in inches of the closest object detected by the distance finder.

   ```
   float DistInches = 0;
   ```

5. Initializes to 0 the `DistTrip` variable, which holds the maximum distance from the sensor at which an object will activate the alarm. This value is calculated every time the alarm is reset.

   ```
   float DistTrip = 0;
   ```

6. Initializes to 5 inches the `DistDelta` variable, which holds the offset distance that is used in determining the trip distance.

   ```
   float DistDelta = 5;
   ```

7. Assigns to pin 9 on the Arduino the `BuzzerPin` variable, which represents the positive terminal on the buzzer.

   ```
   int BuzzerPin = 9;
   ```

8. Initializes to 300 the `BuzzerFreq` variable, which holds the frequency of the alarm tone

generated by the buzzer when the alarm system has been tripped.

```
int BuzzerFreq = 300;
```

9. Initializes to 600 the `BuzzerFreqWait` variable, which holds the frequency of the wait state tone generated by the buzzer when the alarm system is in the "Waiting" state.

```
int BuzzerFreqWait = 600;
```

10. Assigns to pin 6 on the Arduino the `ButtonPin` variable, which represents the reset button.

```
int ButtonPin = 6;
```

11. Initializes to 0 tThe `RawValue` variable, which holds the value read from the Reset button.

```
int RawValue = 0;
```

12. Enumerates the `AlarmState` variable, which defines the states of the alarm system as "Waiting," "Active," or "Tripped."

```
enum AlarmState
{
Waiting,
Active,
Tripped
};
```

13. Initializes the `CurrentAlarmState` variable, which holds the current state of the alarm system, to "Waiting."

```
AlarmState CurrentAlarmState =
Waiting;
```

14. Calls up the `setup()` function, which:

   a. Sets the pin on the Arduino that is connected to the Trigger input pin on the distance sensor to be an output pin so that voltages can be delivered to the sensor.

   b. Sets the pin on the Arduino that is connected to the Echo output pin on the distance sensor to be an input

pin so that data can be read from the sensor.

   c. Sets the pin on the Arduino that is used to read the status of the alarm's Reset button to be an input pin.

   d. Initializes the Serial Monitor, and sets the communication speed to 9,600 baud.

   e. Prints a text message to the Serial Monitor indicating that the program is starting.

   f. Calculates the trip distance by calling the `SetDistanceTrip()` function.

See Listing 7-5.

Listing 7-5 The `setup()` Function

```
void setup()
{
 pinMode(TriggerPin, OUTPUT);
 pinMode(EchoPin, INPUT);
 pinMode(ButtonPin, INPUT);

 Serial.begin(9600);
 Serial.println("HC-SR04 Distance
   Sensor Alarm ....");

 SetDistanceTrip();
}
```

15. Calls up the `ProcessWaitingState()` function, which processes the "Waiting" alarm state by:

   a. Generating 15 short beeps using the buzzer.

   b. Generating 1 long beep using the buzzer.

   c. Changing the alarm system state to "Active."

   d. Printing out a text message to the Serial Monitor indicating the current state of the alarm.

See Listing 7-6.

Listing 7-6 The `ProcessWaitingState()` Function

```
void ProcessWaitingState()
{
 for(int i = 0; i < 15; i++)
 {
    // Play Wait Tone
    tone(BuzzerPin, BuzzerFreqWait);
    delay (300);

    noTone(BuzzerPin);
    delay (300);
 }

// Final Beep
tone(BuzzerPin, BuzzerFreqWait);
delay(1500);

CurrentAlarmState = Active;
Serial.println("CurrentAlarmState:
   Active");
}
```

16. Calls up the `MicrosecondsToInches()` function, which converts the duration of the echo pulse to a distance in inches. This distance is measured from the sensor to the object that is in front of the sensor. See Listing 7-7.

Listing 7-7 The `MicrosecondsToInches()` Function

```
float MicrosecondsToInches(long
   Microseconds)
{
 // Distance in inches = uS/148
 return Microseconds/148.0;
}
```

17. Calls up the `SendTriggerSignal()` function, which sends a trigger signal to the distance sensor that activates a series of sonic pulses by:

    a. Writing a 0 or LOW value to the trigger input on the distance sensor.

    b. Suspending execution of the program for 2 microseconds.

    c. Writing a 1 or HIGH value to the trigger input on the distance sensor.

    d. Suspending execution of the program for 10 microseconds.

    e. Writing a 0 or LOW to the trigger input on the distance sensor.

    See Listing 7-8.

Listing 7-8 The `SendTriggerSignal()` Function

```
void SendTriggerSignal()
{
 // Send Sound Pulse to detect object
 digitalWrite(TriggerPin, 0);
 delayMicroseconds(2);
 digitalWrite(TriggerPin, 1);
 delayMicroseconds(10);
 digitalWrite(TriggerPin, 0);
}
```

18. Calls up the `GetDistanceInches()` function, which retrieves the distance between the sensor and an object in front of the sensor by:

    a. Sending a trigger signal to the distance sensor to initiate a series of sonic pulses by calling the `SendTriggerSignal()` function.

    b. Reading the returned echo pulse data from the distance sensor and determining the duration of the pulse in microseconds.

    c. Converting the duration of the echo pulse from step b to a distance in inches by calling the `MicrosecondsToInches (DurationPulse)` function.

    d. Returning the distance from step c.

    See Listing 7-9.

Listing 7-9 The `GetDistanceInches()` Function

```
float GetDistanceInches()
{
 float retval = 0;

 SendTriggerSignal();

 // Read in return echo data
 DurationPulse = pulseIn(EchoPin, HIGH);

 // Convert the returned echo pulse into
 // a distance
 retval = MicrosecondsToInches(Duration
   Pulse);

 return retval;
}
```

19. Calls up the `SetDistanceTrip()` function, which sets the trip distance by:

    a. Determining the distance to the object in front of the range finder without an intruder present by calling the `GetDistanceInches()` function.

    b. Determining the trip distance by subtracting `DistDelta` from the distance determined in step a.

    c. Printing out the trip distance to the Serial Monitor.

    See Listing 7-10.

Listing 7-10 The `SetDistanceTrip()` Function

```
void SetDistanceTrip()
{
 // Get current distance without object
 DistTrip = GetDistanceInches();
 DistTrip = DistTrip - DistDelta;

 // Print out default distance without
 // objects
 Serial.print("DistTrip: ");
 Serial.println(DistTrip);
}
```

20. Calls up the `loop()` function, which contains the main alarm system logic and:

    a. Reads in the status of the Reset button.

    b. If the Reset button is being pressed, the function

       i. Sets the maximum distance at which the alarm will be tripped by calling the `SetDistanceTrip()` function.

       ii. Changes the current alarm system status to "Waiting."

       iii. Prints out the status of the alarm system to the Serial Monitor.

    c. If the alarm status is "Waiting," then the function processes this state by calling the `ProcessWaitingState()` function.

    d. If the alarm status is "Active," the function:

       i. Finds the current distance between the sensor and the object in front of the sensor by calling the `GetDistanceInches()` function.

       ii. If the current distance is less than the alarm trip distance, the function changes the alarm state to "Tripped" to indicate that an intruder has been detected.

       iii. Prints out to the Serial Monitor that an intruder has been detected and at what distance from the sensor it was detected at.

       iv. Prints out to the Serial Monitor the updated status of the alarm system.

    e. If the alarm status is "Tripped," then the function generates the alarm tone using the buzzer. Otherwise, it makes sure that any sounds produced by the buzzer are stopped.

    See Listing 7-11.

Listing 7-11   The `loop()` Function

```
void loop()
{
 // Read Reset Button
 RawValue = digitalRead(ButtonPin);
 if (RawValue)
 {
    // Set Distance at which alarm will be tripped
    SetDistanceTrip();

    // Set new alarm status to waiting
    CurrentAlarmState = Waiting;
    Serial.println("CurrentAlarmState: Waiting");
 }

 // Check if Alarm is in Waiting State
 if (CurrentAlarmState == Waiting)
 {
    ProcessWaitingState();
 }

 // Trigger and read distance sensor if alarm active
 if (CurrentAlarmState == Active)
 {
    // Read current distance
    DistInches = GetDistanceInches();
    if (DistInches < DistTrip)
    {
       // Object detected
       CurrentAlarmState = Tripped;
       Serial.println("Object Detected .... ");
       Serial.print("Object Distance (inches): ");
       Serial.println(DistInches);
       Serial.println("CurrentAlarmState: Tripped");
    }
 }

 // Sound Alarm if alarm has been tripped.
 if (CurrentAlarmState == Tripped)
 {
    tone(BuzzerPin, BuzzerFreq);
 }
 else
 {
    noTone(BuzzerPin);
 }
}
```

Running the Program

The distance sensor works best when pointed at a flat surface such as a door with no objects obstructing the path between the sensor and the flat surface. For example, try placing the sensor on the edge of a table and pointing it at a door. Upload the program to the Arduino, and start up the Serial Monitor. You should see the startup message and the trip distance displayed.

```
HC-SR04 Distance Sensor Alarm ....
DistTrip: 35.18
```

The alarm should be in the "Waiting" state, and you should hear a series of short beeps followed by a long beep. The alarm system is now active.

```
CurrentAlarmState: Active
```

Trigger the alarm by passing your hand in front of the sensor. The distance that triggered the alarm and the current state of the alarm should be printed out to the Serial Monitor.

```
Object Detected ....
Object Distance (inches): 6.32
CurrentAlarmState: Tripped
```

You should hear a constant alarm tone from the buzzer. Now press the Reset button. You should see the new calculated trip distance displayed, and a series of short beeps should start to be produced and the state of the alarm changed to "Waiting."

```
DistTrip: 34.71
CurrentAlarmState: Waiting
```

When you hear a long beep, the alarm system will then change state to "Active," where intruders can be detected.

```
CurrentAlarmState: Active
```

Hands-on Example: Arduino Collision Alarm

In this example project, you will build an Arduino collision alarm system. This system uses a buzzer to alert the user to the distance between the sensor and the object in front of the sensor. The buzzer emits a series of beeps if the object is within a certain range. The beeps increase in tone and frequency the closer the object gets to the sensor. If no object is detected within the target distance range, then the buzzer doesn't sound.

Parts List

To build this example project, you will need

- 1 HC-SR04 distance sensor
- 1 piezo buzzer
- 1 breadboard (recommended to hold the distance sensor)
- Wires to connect the components to the Arduino

Setting Up the Hardware

To build the hardware for this example project, you will need to:

- Connect the GND pin on the distance sensor to a GND pin on the Arduino.
- Connect the Trigger pin on the distance sensor to pin 8 on the Arduino.
- Connect the Echo pin on the distance sensor to pin 7 on the Arduino.
- Connect the VCC pin on the distance sensor to the 5-V pin on the Arduino.
- Connect the negative terminal on the buzzer to a GND pin on the Arduino.
- Connect the positive terminal on the buzzer to pin 9 on the Arduino (Figure 7-11).

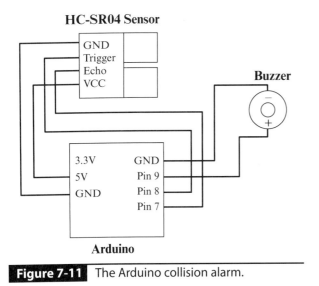

HC-SR04 Sensor

GND
Trigger
Echo
VCC

Buzzer

3.3V   GND
5V   Pin 9
GND   Pin 8
              Pin 7

Arduino

Figure 7-11 The Arduino collision alarm.

Setting Up the Software

The program controls the collision alarm logic and operates the hardware by:

1. Assigning to pin 8 on the Arduino the `TriggerPin` variable, which represents the Trigger pin on the sensor.

   ```
   int TriggerPin = 8;
   ```

2. Assigning to pin 7 on the Arduino the `EchoPin` variable, which represents the Echo pin on the sensor.

   ```
   int EchoPin = 7;
   ```

3. Setting the `DurationPulse` variable, which represents the duration of the echo pulse, which is proportional to the distance an object is from the sensor.

   ```
   long DurationPulse = 0;
   ```

4. Initializing to 0 the `DistInches` variable, which holds the sensor's reading of the distance to the nearest object detected by the sensor in inches.

   ```
   float DistInches = 0;
   ```

5. Setting the `DistFar` variable, which holds the greatest distance from the sensor in inches that an object will be considered to be in the collision range, where the buzzer will start beeping to indicate a collision. The default value is 15 inches.

   ```
   float DistFar = 15;
   ```

6. Setting the `DistNear` variable, which holds the distance where the beeps produced by the buzzer will be the highest pitched and most frequent. This default value of this variable is 3 inches.

   ```
   float DistNear = 3;
   ```

7. Assigning to pin 9 on the Arduino the `BuzzerPin` variable, which represents the positive terminal of the buzzer.

   ```
   int BuzzerPin = 9;
   ```

8. Setting the `BuzzerFreqLow` variable, which holds the lowest tone that the buzzer will generate when the sensor detects an object in the collision area. This lowest tone is generated when the sensor detects an object at `DistFar` distance from the sensor. The default value of this variable is 100.

   ```
   int BuzzerFreqLow = 100;
   ```

9. Setting the `BuzzerFreqHigh` variable, which represents the highest frequency generated by the buzzer when an object is in the collision range. This frequency starts to be generated when an object is at or less than `DistNear` inches from the sensor. The default value is set to 800.

   ```
   int BuzzerFreqHigh = 800;
   ```

10. Setting the `BuzzerTone` variable, which holds the calculated tone of the beeps that will be generated when an object is detected by the sensor in the collision area.

    ```
    int BuzzerTone = 0;
    ```

11. Setting the `BuzzerDelayMin` variable, which holds the time in milliseconds between beeps made by the buzzer when the sensor detects an object within `DistNear` inches of the sensor. The time between beeps is initialized to 0.

    ```
    int BuzzerDelayMin = 0;
    ```

12. Setting the `BuzzerDelayMax` variable, which holds the time in milliseconds between beeps made by the buzzer when the sensor detects an object `DistFar` inches away. The time is initialized to 1,000 milliseconds, or 1 second, between beeps.

    ```
    int BuzzerDelayMax = 1000;
    ```

13. Setting the `BuzzerDelay` variable, which holds the actual calculated delay between beeps produced when an object is detected within `DistFar` inches of the sensor.

    ```
    int BuzzerDelay = 0;
    ```

14. Setting the `BuzzerDelayStartTime` variable, which holds the time that a beep from the buzzer has just been completed and is used in determining if `BuzzerDelay` milliseconds have passed so that the next beep can be emitted from the buzzer.

    ```
    unsigned long BuzzerDelayStartTime = 0;
    ```

15. Calling the `setup()` function, which:

    a. Sets the pin on the Arduino that is connected to the Trigger input pin on the distance sensor to be an output pin so that voltages can be delivered to the sensor.

    b. Sets the pin on the Arduino that is connected to the Echo output pin on the distance sensor to be an input pin so that data can be read from the sensor.

    c. Initializes the Serial Monitor and sets the communication speed to 9,600 baud.

    d. Prints out a text message to the Serial Monitor indicating that the program is starting.

    e. Suspends the program for 100 microseconds to allow for the distance sensor to initialize.

    See Listing 7-12.

Listing 7-12 The `setup()` Function

```
void setup()
{
  pinMode(TriggerPin, OUTPUT);
  pinMode(EchoPin, INPUT);

  Serial.begin(9600);
  Serial.println("HC-SR04 Distance Sensor
    Collision Alarm ....");

  delay(100);
}
```

16. Calling the `MicrosecondsToInches()` function, which converts the duration of the echo pulse to a distance in inches. This distance is measured from the sensor to the object that is in front of the sensor.

    See Listing 7-13.

Listing 7-13 The `MicrosecondsToInches()` Function

```
float MicrosecondsToInches(long
Microseconds)
{
  // Distance in inches = uS/148
  return Microseconds/148.0;
}
```

17. Calling the `SendTriggerSignal()` function, which sends a trigger signal to the distance sensor that activates a series of sonic pulses by:

    a. Writing a 0 or LOW value to the Trigger input on the distance sensor.

    b. Suspending execution of the program for 2 microseconds.

c. Writing a 1 or HIGH value to the Trigger input on the distance sensor.

d. Suspending execution of the program for 10 microseconds.

e. Writing a 0 or LOW value to the Trigger input on the distance sensor.

See Listing 7-14.

Listing 7-14 The `SendTriggerSignal()` Function

```
void SendTriggerSignal()
{
  // Send Sound Pulse to detect object
  digitalWrite(TriggerPin, 0);
  delayMicroseconds(2);
  digitalWrite(TriggerPin, 1);
  delayMicroseconds(10);
  digitalWrite(TriggerPin, 0);
}
```

18. Calling the `GetDistanceInches()` function, which retrieves the distance between the sensor and an object in front of the sensor by:

a. Sending a trigger signal to the distance sensor to initiate a series of sonic pulses by calling the `SendTriggerSignal()` function.

b. Reading the returned echo pulse data from the distance sensor and determining the duration of the pulse in microseconds.

c. Converting the duration of the echo pulse from step b to a distance in inches by calling the `MicrosecondsToInches(DurationPulse)` function.

d. Returning the distance from step c.

See Listing 7-15.

Listing 7-15 The `GetDistanceInches()` Function

```
float GetDistanceInches()
{
  float retval = 0;

  // Trigger sensor's sound pulses
  SendTriggerSignal();

  // Read in return echo data
  DurationPulse = pulseIn(EchoPin, HIGH);

  // Convert the returned echo pulse into
  // a distance
  retval = MicrosecondsToInches(Duration
    Pulse);

  return retval;
}
```

19. Calling the `UpdateBuzzer()` function, which updates the beeps produced by the buzzer when an object has been detected within the collision area of the sensor by:

a. Calculating the delay (in microseconds) between beeps by finding the relative position of the object within the range of `DistNear` and `DistFar` and then mapping this to the same position within the range of `BuzzerDelayMin` and `BuzzerDelayMax` by using the built-in `map()` function.

b. Limiting the value of the delay between `BuzzerDelayMin` and `BuzzerDelayMax`.

c. Calculating the elapsed time since the last beep was emitted by the buzzer. If the elapsed time is less than the buzzer delay time, then the function terminates.

d. Calculating the pitch of the beep by finding the relative position of the detected object within the range of `DistNear` and `DistFar` and then finding the same relative position within the range of `BuzzerFreqHigh` and `BuzzerFreqLow`. For example, when the object is at `DistFar` distance from the sensor, the pitch of the beeps would be `BuzzerFreqLow`, which would produce the lowest pitch.

e. Limiting the pitch of the beep from step d to between `BuzzerFreqLow` and `BuzzerFreqHigh` inclusive. If the distance of the detected object is less than or equal to `DistFar`, which is the limit of the collision area, then the function produces 1 beep using the pitch calculated in step d. Otherwise, if the detected object is outside the collision area, then the function stops any sounds being produced by the buzzer.

See Listing 7-16.

Listing 7-16 The `UpdateBuzzer()` Function

```
void UpdateBuzzer()
{
 unsigned long ElapsedTime = 0;

 // Update Buzzer Delay
 BuzzerDelay = map(DistInches, DistNear, DistFar, BuzzerDelayMin, BuzzerDelayMax);
 BuzzerDelay = constrain(BuzzerDelay, BuzzerDelayMin, BuzzerDelayMax);

 ElapsedTime = millis() - BuzzerDelayStartTime;
 if (ElapsedTime < BuzzerDelay)
 {
    return;
 }

 // Map distance to buzzer tone
 BuzzerTone = map(DistInches, DistNear, DistFar, BuzzerFreqHigh, BuzzerFreqLow);
 BuzzerTone = constrain(BuzzerTone, BuzzerFreqLow, BuzzerFreqHigh);

 if (DistInches <= DistFar)
 {
    // Produce Beep
    tone(BuzzerPin, BuzzerTone);
    delay(300);
    noTone(BuzzerPin);
    BuzzerDelayStartTime = millis();
 }
 else
 {
    // No tone
    noTone(BuzzerPin);
 }
}
```

20. Calling the `PrintDebugInfo()` function, which prints out the distance in inches of the object that is detected by the sensor and the pitch of the buzzer to the Serial Monitor.

See Listing 7-17.

Listing 7-17 The `PrintDebugInfo()` Function

```
void PrintDebugInfo()
{
 // Distance
 Serial.print("Distance: ");
 Serial.print(DistInches);

 // Buzzer Tone
 Serial.print(" , BuzzerTone: ");
 Serial.println(BuzzerTone);
}
```

21. Executing the `loop()` function continuously, which executes the main code for the collision alarm system by:

   a. Reading the distance sensor and returning the distance of the object that is in front of the sensor in inches by calling the `GetDistanceInches()` function.

   b. Updating the status of any beeps that need to be produced if the detected object is within the collision area by calling the `UpdateBuzzer()` function.

   c. If the detected object is within the object collision area, then the function prints out some debug information by calling the `PrintDebugInfo()` function.

   See Listing 7-18.

Listing 7-18 The `loop()` Function

```
void loop()
{
 // Get Distance to Object in front of
 // sensor
 DistInches = GetDistanceInches();

 // Update Buzzer
 UpdateBuzzer();

 // Print Collision Alarm Debug
 // information
 if (DistInches <= DistFar)
 {
     PrintDebugInfo();
 }
}
```

Running the Program

Upload the program to the Arduino, and start the Serial Monitor. In a few moments you should see the program initialization message.

```
HC-SR04 Distance Sensor Collision Alarm
. . . .
```

Put your hand in front of the distance sensor, and you should hear beeping from the buzzer and messages on the Serial Monitor that display the distance of your hand from the sensor and the pitch of the buzzer.

```
Distance: 14.26 , BuzzerTone: 159
```

Move your hand toward the sensor. You should hear the beeps grow in frequency and pitch. When you get at `DistNear` distance from the sensor, the frequency and pitch of the beeps should be at their maximum. Move your hand away from the sensor, and the beeps should decrease in frequency and pitch.

Summary

This chapter has covered sensors that can be used to detect the presence of humans. It began by presenting a glass break alarm system for the Arduino and the Raspberry Pi. This was followed by a description of an infrared motion detector called the HC-SR501. Then we built infrared motion alarm systems using this sensor for the Arduino and the Raspberry Pi. This was followed by a discussion of the HC-SR04 distance sensor. Then we built an intruder alarm system using this sensor for the Arduino. Finally, we built an Arduino collision alarm system using the HC-SR04.

## CHAPTER 8

Arduino TFT LCD Display and Image Sensor Projects

THIS CHAPTER DISCUSSES the Arduino TFT LCD and image sensors. First, we introduce the thin-film-transistor (TFT) liquid-crystal display (LCD) with SD card reader/writer that is specifically designed for the Arduino. Then a hands-on example shows you how to use the TFT display and the attached SD card reader/writer. You will learn how to draw text and animated lines, circles, rectangles, and points to the TFT display. You also learn how to use the SD card to write and read data from an SD card. Then another hands-on example shows you how to use the TFT display and SD card reader/writer as part of a door alarm system that also tracks the number of times the door alarm has been tripped. Finally, we discuss cameras that can be used with the Arduino and cover the features of the ArduCAM OV2640 minicamera. The chapter concludes with a hands-on example using the AduCAM minicamera that shows you how to take a photo and then save it to an SD card by pressing a button.

Arduino TFT LCD Color Display Screen with SD Card Reader/Writer

The TFT display screen with SD card reader and writer is a standard device that is supported natively in the Arduino Integrated Development Environment (IDE) version 1.0.5 and above. The built-in Arduino library provides for the display of color text, images, and shapes such as rectangles, circles, lines, and points using native built-in functions. In addition, the TFT display device also contains an SD card reader and writer that can be used to read and write data from a small micro-SD card. The SD card slot is on the back on the display near the top and middle of the device (Figure 8-1).

The pin connections to the Arduino board depend on the specific model of the Arduino board you are using. The following pin connections are for the Arduino Uno. For example, the 5-V pin on the TFT display connects to the 5-V pin of the Arduino. The MISO pin on the TFT board connects to pin 12 of the Arduino board:

+5V	+5V
MISO	pin 12
SCK	pin 13
MOSI	pin 11
LCD CS	pin 10
SD CS	pin 4
D/C	pin 9
RESET	pin 8
BL	+5V
GND	GND

Figure 8-1 TFT display with SD card reader/writer for Arduino.

The recommended way to connect this board to the Arduino is to use a breadboard. The TFT display is inserted into the breadboard, and wires connect each of the pins on the TFT display to the pins on the Arduino. Figure 8-2 shows how to connect the TFT display to the Arduino Uno. The TFT display is facing up with the SD card on top and the pins that need to be connected to the Arduino on the right-hand side.

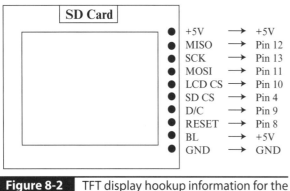

Figure 8-2 TFT display hookup information for the Arduino Uno.

Hands-on Example: Arduino TFT Display and SD Card Reader/Writer Test

In this hands-on example project, you learn how to connect and use a standard TFT color LCD display with SD card reader/writer with an Arduino Uno. The TFT display connections described in this section are specific to the Arduino Uno. In this example project, you will learn how to display text, lines, rectangles, circles, and points on the TFT screen. In addition, you will learn how to read and write data using the SD card that is built into the back of the TFT LCD display device.

Parts List

To build this on example project, you will need

- 1 Arduino TFT LCD display with built-in SD card reader/writer
- 1 micro-SD card

- 1 breadboard (needed to hold the TFT LCD display)

- 1 Arduino Uno (Connection instructions in this example project are specifically designed for the Arduino Uno.)

- Wires to connect the TFT LCD display to the Arduino

Setting Up the Hardware

To connect the hardware for this hands-on example project, you will need to:

- Connect the 5-V pin on the TFT display to the 5-V pin of the Arduino Uno.

- Connect the MISO pin on the TFT display to pin 12 of the Arduino Uno.

- Connect the SCK pin on the TFT display to pin 13 of the Arduino Uno.

- Connect the MOSI pin on the TFT display to pin 11 of the Arduino Uno.

- Connect the LCD CS pin on the TFT display to pin 10 of the Arduino Uno.

- Connect the SD CS pin on the TFT display to pin 4 of the Arduino Uno.

- Connect the D/C pin on the TFT display to pin 9 of the Arduino Uno.

- Connect the RESET pin on the TFT display to pin 8 of the Arduino Uno.

- Connect the BL pin on the TFT display to the 5-V pin of the Arduino Uno.

- Connect the GND pin on the TFT Display to the GND pin of the Arduino Uno (Figure 8-3).

Setting Up the Software

The software for this example project draws text and shapes and displays images from the SD card onto the TFT LCD display. It also reads

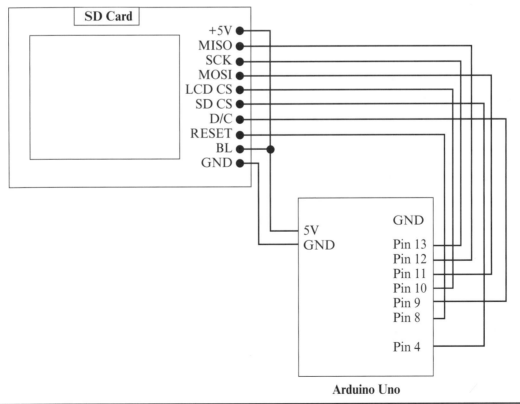

Figure 8-3   Arduino TFT display and SD card reader/writer test.

and writes test data using the SD card. The program does this by:

1. Including the Serial Peripheral Interface (SPI) library. This library is used to communicate with the TFT display and the SD card.

   `#include <SPI.h>`

2. Including the SD library. The SD library is used to provide functions for the reading and writing of data to the SD card that is built into the TFT display.

   `#include <SD.h>`

3. Including the TFT library. This is done so that functions that control the TFT display can be accessed.

   `#include <TFT.h>`

4. Defining the chip select pin for the SD card as pin 4 of the Arduino. If the chip select pin is set to 0 or low voltage, the SD card communicates with the Arduino. If the chip select pin is set to 1 or high voltage, the SD card ignores the Arduino.

   `#define sd_cs 4`

5. Defining the chip select pin for the TFT LCD screen as pin 10 of the Arduino:

   `#define lcd_cs 10`

6. Defining the DC pin on the TFT display as connected to pin 9 of the Arduino:

   `#define dc 9`

7. Defining the Reset pin on the TFT display as connected to pin 8 of the Arduino:

   `#define rst 8`

8. Defining the variable `TFTScreen`, which is an object of type TFT that represents the TFT display screen:

   `TFT TFTScreen = TFT(lcd_cs, dc, rst);`

9. Defining the `BitmapImage` variable, which holds the actual bitmap image that is loaded into and displayed on the TFT screen using

`PImage` class. The `PImage` class is part of the standard TFT library.

`PImage BitmapImage;`

10. Using the `TestFile` variable to read and write data to a file employing the SD card reader/writer located on the back of the TFT display:

    `File TestFile;`

11. Initializing the `setup()` function, which:

    a. Initializes the Serial Monitor with a communication speed of 9,600 baud.

    b. Printing a text message to the Serial Monitor indicating the start of the program.

    c. Initializes the TFT screen device.

    d. Setting the background color of the TFT screen to black by calling the `TFTScreen.background(0, 0, 0)` function with all zeros for all the color input values.

    e. Setting the color of the text that will be placed on the TFT screen to white by calling the `TFTScreen.stroke(255, 255, 255)` function with all 255s, which is full intensity for all color input values.

    f. Printing some text onto the TFT screen indicating that the program has started.

    g. Delaying program execution for 3 seconds.

    h. Initializing the SD card reader and writer.

    i. Printing the status of the SD card reader/writer initialization to the Serial Monitor. See Listing 8-1.

The `LoadImage()` function:

1. Loads in a bitmap image called `filename` from the SD card that is attached to the TFT display. The filename is an input character array to the function.

Listing 8-1 The `setup()` Function

```
void setup()
{
 Serial.begin(9600);
 Serial.println(F("TFT Display and SD Card Test ..."));

 TFTScreen.begin();
 TFTScreen.background(0, 0, 0);
 TFTScreen.stroke(255, 255, 255);
 TFTScreen.println(F("Arduino TFT Display/SD Card Example"));
 delay(3000);

 Serial.print(F("Initializing SD card..."));
 if (!SD.begin(sd_cs))
 {
    Serial.println(F("SD Card Initialization Failed ..."));
    return;
 }
 Serial.println(F("SD Card Initialization OK ..."));
}
```

2. Checks to see whether the bitmap image is valid.

3. If the bitmap image is not valid, prints a text error message to the Serial Monitor.

4. Displays the bitmap image on the TFT screen starting at the upper left-hand corner of the display. See Listing 8-2.

Listing 8-2 The `LoadImage()` Function

```
void LoadImage(char* filename)
{
 BitmapImage = TFTScreen.
   loadImage(filename);
 if (!BitmapImage.isValid())
 {
    Serial.print(F("Error While Loading
      in "));
    Serial.println(filename);
 }
 TFTScreen.image(BitmapImage, 0, 0);
}
```

The `LoadTestImage()` function:

1. Clears the TFT background screen to black.

2. Loads and displays the bitmap on the SD card attached to the TFT display with the filename `testbmp.bmp`.

3. Suspends execution of the program for 3 seconds. See Listing 8-3.

Listing 8-3 The `LoadTestImage()` Function

```
void LoadTestImage()
{
 // Loading bitmap from SD Card
 TFTScreen.background(0, 0, 0);
 LoadImage("testbmp.bmp");
 delay(3000);
}
```

The `DrawLines()` function animates a line on the TFT screen by:

1. Clearing the TFT screen and setting the background color to black.

2. Drawing a series of white lines across the TFT screen that move from left to right by:

a.   Setting the line color to white.

b.   Drawing a line starting at x position 0 and y position 10 on the TFT screen and ending at x position i and y position 60. The value of i ranges from 0 to the width of the TFT screen in pixels.

c.   Setting the line color to black.

d.   Drawing a line starting from x position 0 and y position 10 on the TFT screen and ending at x position i and y position 60. The range of the values for i is from 0 to the width of the TFT screen in pixels. By drawing a black line over the white line that was just drawn, you are erasing the white line and performing a simple form of animation.

3.   Drawing a series of white lines across the TFT screen that move from right to left by:

a.   Setting the color of the line to white.

b.   Drawing a line starting at x position 0 and y position 10 on the TFT screen and ending at x position i and y position 60. The value of i ranges from the width of the TFT screen in pixels to 1.

c.   Setting the color of the line to black.

d.   Drawing a black line over the white line.

4.   Suspending execution of the program for 3 seconds. See Listing 8-4.

The `DrawRectangle()` function moves a blue rectangle across the screen of the TFT by:

1.   Clearing the background of the TFT and setting the background color to black.

2.   Turning off the outline color of the shapes that are drawn.

Listing 8-4   The `DrawLines()` Function

```
void DrawLines()
{
  // Drawing line to screen
  TFTScreen.background(0, 0, 0);
  for (int i = 0; i < TFTScreen.width();
    i++)
  {
     // Draw Line
     TFTScreen.stroke(255, 255, 255);
     TFTScreen.line(0, 10, i, 60);

     // Erase Line
     TFTScreen.stroke(0, 0, 0);
     TFTScreen.line(0, 10, i, 60);
  }
  for (int i = TFTScreen.width(); i > 0;
    i--)
  {
     // Draw Line
     TFTScreen.stroke(255, 255, 255);
     TFTScreen.line(0, 10, i, 60);

     // Erase Line
     TFTScreen.stroke(0, 0, 0);
     TFTScreen.line(0, 10, i, 60);
  }
  delay(3000);
}
```

3.   Drawing a rectangle that moves from the left-hand side of the TFT screen to the right-hand side of the screen by:

a.   Setting the rectangle color to blue.

b.   Drawing the rectangle with the upper left-hand corner located at (i,10) with a width of 40 pixels and a height of 50 pixels. The i value will range from 0 to 40 pixels less than the width of the screen. By doing this, the entire rectangle will stay within the full view of the screen.

c.   Setting the rectangle color to black.

d. Drawing the rectangle at the same location as in step b, which erases the rectangle from the TFT display.

4. Drawing a rectangle that moves from the right-hand side to the left-hand side of the TFT screen by:

   a. Setting the rectangle color to blue.

   b. Drawing the rectangle with a width of 40 pixels and a height of 50 pixels at an *x* position of *i* and a *y* position of 10. The value of *i* will range from 40 less than the width of the TFT screen to 1.

c. Setting the rectangle color to black.

d. Drawing the rectangle at the same position as in step b, which will have the effect of erasing that rectangle.

5. Setting the rectangle color to blue.

6. Drawing a rectangle of width 40 pixels and height of 50 pixels at location (0,10).

7. Delaying program execution for 3 seconds. See Listing 8-5.

Listing 8-5 The `DrawRectangle()` Function

```
void DrawRectangle()
{
 TFTScreen.background(0, 0, 0);
 TFTScreen.noStroke();

 // screen.rect(xStart, yStart, width, height);
 // Move rectangle to right
 for (int i = 0; i < (TFTScreen.width()-40); i++)
 {
    // Draw
    TFTScreen.fill(255,0,0);
    TFTScreen.rect(i,10,40,50);

    // Erase
    TFTScreen.fill(0,0,0);
    TFTScreen.rect(i,10,40,50);
 }

 // Move rectangle to left
 for (int i = (TFTScreen.width()-40); i > 0; i--)
 {
    // Draw
    TFTScreen.fill(255,0,0);
    TFTScreen.rect(i,10,40,50);

    // Erase
    TFTScreen.fill(0,0,0);
    TFTScreen.rect(i,10,40,50);
 }

 // Draw Final Position
 TFTScreen.fill(255,0,0);
 TFTScreen.rect(0,10,40,50);

 delay(3000);
}
```

The `DrawCircle()` function draws and animates a circle on the TFT screen by:

1. Clearing the background of the TFT screen and setting the background color to black.

2. Disabling the drawing of the outline color for the circle by calling the `noStroke()` function.

3. Drawing a series of circles on the TFT screen that will move the circle from the left-hand side of the screen to the right-hand side of the screen by:

    a. Setting the color of the circle to green.

    b. Drawing the circle at *x* and *y* position (*i*,20) with a radius of 20 pixels. The *i* value will range from 20 to 40 less than the width of the TFT screen.

    c. Changing the color of the circle to black.

    d. Drawing the circle at the same location as in step b, which has the effect of erasing the circle on the TFT display.

4. Drawing a series of circles on the TFT screen that will move the circle from the right-hand side of the screen to the left-hand side by:

    a. Setting the color of the circle to green.

    b. Drawing the circle at *x* and *y* position (*i*,20) with a radius of 20 pixels. The range of the *i* values will be from 40 less than the width of the LCD screen to 21. This moves the circle from the right-hand side of the screen to the left-hand side.

    c. Setting the color of the circle to black.

    d. Draw the circle at the same position as in step b, which has the effect of erasing the previously drawn circle.

5. Setting the color of the circle to green.

6. Drawing the circle at *x,y* location (20,20) with a radius of 20 pixels.

7. Suspending program execution for 3 seconds. See Listing 8-6.

Listing 8-6   The `DrawCircle()` Function

```
void DrawCircle()
{
 TFTScreen.background(0, 0, 0);
 TFTScreen.noStroke();

 //screen.circle(xPos, yPos, radius);
 // Move circle right
 for (int i = 20; i <
   (TFTScreen.width()-40); i++)
 {
    // Draw
    TFTScreen.fill(0,255,0);
    TFTScreen.circle(i,20,20);

    // Erase
    TFTScreen.fill(0,0,0);
    TFTScreen.circle(i,20,20);
 }

 // Move circle to the left
 for (int i = (TFTScreen.width()-40);
   i > 20; i--)
 {
    // Draw
    TFTScreen.fill(0,255,0);
    TFTScreen.circle(i,20,20);

    // Erase
    TFTScreen.fill(0,0,0);
    TFTScreen.circle(i,20,20);
 }
 // Final Position
 TFTScreen.fill(0,255,0);
 TFTScreen.circle(20,20,20);

 delay(3000);
}
```

The `DrawPoint()` function draws points to the TFT screen by:

1. Clearing the TFT screen and setting the background color to black.

2. Setting the color of the point to white by calling the `stroke()` function with 255 entered for the blue, green, and red parameters.

3. Drawing a line that moves from the left-hand side of the TFT screen to the right-hand side by repeatedly:

   a. Changing the color of the point to white.

   b. Drawing a point to the TFT screen at *x,y* position (*i*,20). The variable *i* will range from 0 to the width of the TFT screen.

   c. Suspending program execution for 50 milliseconds.

4. Erasing the line that was drawn in step 3 by repeatedly:

   a. Changing the color of the point to black.

   b. Drawing a point to the TFT screen at *x,y* location (*i*,20). The value of *i* will range from the width of the TFT screen to 1. This will have the effect of erasing the white line drawn in step 3 and making the line appear to move back to the left of the screen.

   c. Suspending execution of the program for 50 milliseconds.

5. Suspending execution of the program for 3 seconds. See Listing 8-7.

Listing 8-7   The `DrawPoint()` Function

```
void DrawPoint()
{
 TFTScreen.background(0, 0, 0);
 TFTScreen.stroke(255,255,255);

 // Move line to right
 for (int i = 0; i < TFTScreen.width();
   i++)
 {
   // Draw
   TFTScreen.stroke(255,255,255);
   TFTScreen.point(i,20);
   delay(50);
 }

 // Move Line to left
 for (int i = TFTScreen.width(); i > 0;
   i--)
 {
   // Draw
   TFTScreen.stroke(0,0,0);
   TFTScreen.point(i,20);
   delay(50);
 }
 delay(3000);
}
```

The `WriteFile()` function tests the SD card attached to the back of the TFT screen by:

1. Clearing the TFT screen and setting the background color to black.

2. Setting the color of the text on the TFT screen to white.

3. Printing a message that a file is being written to the SD card on the TFT display and to the Serial Monitor.

4. Removing that file and printing the result to the TFT screen and to the Serial Monitor if the file to be written already exits.

5. Opening the file for writing.

6. If the file has been opened successfully for writing, then:

   a. Writing text and data to the opened file.

   b. Closing the file.

   c. Printing a text message indicating that the file was written to the TFT screen and to the Serial Monitor.

7. If the file was not opened successfully, then printing out a text message to the TFT and Serial Monitor indicating that there was an error when opening the file for writing.

8. Suspending program execution for 3 seconds. See Listing 8-8.

Listing 8-8 The `WriteFile()` Function

```
void WriteFile(char* filename)
{
 TFTScreen.background(0, 0, 0);
 TFTScreen.stroke(255, 255, 255);
 TFTScreen.println(F("Writing Test File to SD Card ..."));
 Serial.println(F("Writing Test File to SD Card ..."));

 if (SD.exists(filename))
 {
    if (SD.remove(filename))
    {
       TFTScreen.println(F("Removing existing file ..."));
       Serial.println(F("Removing existing file ..."));
    }
    else
    {
       TFTScreen.println(F("ERROR ... Failed to remove existing file ..."));
       Serial.println(F("ERROR ... Failed to remove existing file ..."));
    }
 }

 TestFile = SD.open(filename, FILE_WRITE);
 if (TestFile)
 {
    TestFile.println("Text: The quick brown fox jumped over the lazy dog's back.");
    TestFile.print("Data: ");
    TestFile.println(String(1234));
    TestFile.close();
    TFTScreen.println(F("Writing Done ..."));
    Serial.println(F("Writing Done ..."));
 }
 else
 {
    Serial.println(F("Error Opening File For Write ..."));
    TFTScreen.println(F("Error Opening File For Write..."));
 }
 delay(3000);
}
```

The `ReadFile()` function reads the data from the file written by the `WriteFile()` function and displays these data on the TFT screen and the Serial Monitor by:

1. Clearing the TFT screen and setting the background color to black.

2. Setting the text color to white.

3. Printing text to the TFT screen and the Serial Monitor indicating the test file is about to be read from the SD card.

4. Opening the file on the SD card for reading.

5. If the opening was successful, then while there is more data in the file:

   a. Reading in a byte from the file.

   b. Writing the byte to the Serial Monitor.

   c. Writing the byte to the TFT screen.

6. Closing the file.

7. Printing text error messages to the TFT screen and the Serial Monitor if the file fails to open for reading.

8. Suspending execution of the program for 6 seconds. See Listing 8-9.

The `DisplayEndScreen()` function displays the program end screen by:

1. Clearing the TFT screen and setting the background to black.

Listing 8-9 The `ReadFile()` Function

```
void ReadFile(char* filename)
{
 TFTScreen.background(0, 0, 0);
 TFTScreen.stroke(255, 255, 255);
 TFTScreen.println(F("Reading Test File From SD Card ..."));
 Serial.println(F("Reading Test File From SD Card ..."));

 TestFile = SD.open(filename);
 if (TestFile)
 {
    while (TestFile.available())
    {
       unsigned char DataByte = TestFile.read();
       Serial.write(DataByte);
       TFTScreen.write(DataByte);
    }
    TestFile.close();
 }
 else
 {
    Serial.println(F("Error Opening Test File for Reading ..."));
    TFTScreen.println(F("Error Opening Test File for Reading ..."));
 }
 delay(6000);
}
```

2. Setting the text color to white.

3. Printing the text `The End` on the TFT screen and the Serial Monitor. See Listing 8-10.

Listing 8-10 The `DisplayEndScreen()` Function

```
void DisplayEndScreen()
{
 // End of Demo
 TFTScreen.background(0, 0, 0);
 TFTScreen.stroke(255, 255, 255);
 TFTScreen.println(F("The End ..."));

 Serial.println(F("THE END ..."));
}
```

The `loop()` function contains the main demo code and:

1. Loads the test bitmap image from the SD card and displays the image on the TFT screen by calling the `LoadTestImage()` function.

2. Draws a series of lines across the TFT screen by calling the `DrawLines()` function.

3. Draws a moving rectangle across the TFT screen by calling the `DrawRectangle()` function.

4. Draws a moving circle across the TFT screen by calling the `DrawCircle()` function.

5. Draws a moving line across the TFT screen by calling the `DrawPoint()` function.

6. Writes a file to the SD card called `TestF.txt` by calling the `WriteFile("TestF.txt")` function.

7. Reads and displays the content of the file `TestF.txt` that was written in step 6 by calling the `ReadFile("TestF.txt")` function.

8. Displays the end-of-program message on the TFT by calling the `DisplayEndScreen()` function.

9. Enters an infinite loop that permanently suspends the program. See Listing 8-11.

Listing 8-11 The `loop()` Function

```
void loop()
{
 // Load bitmap image from SD card and
 // draw to screen
 LoadTestImage();

 // Draw a series of lines to the screen
 DrawLines();

 // Draw Moving Rectangle
 DrawRectangle();

 // Draw Moving Circle
 DrawCircle();

 // Draw moving Line using Point
 DrawPoint();

 // Write out test file to SD card
 WriteFile("TestF.txt");

 // Reads in test file from SD card
 ReadFile("TestF.txt");

 // Display End Screen
 DisplayEndScreen();

 // End of Demo
 while(1)
 {
   // Do nothing
 }
}
```

Running the Program

Before starting the program, you will need to select or create a bitmap or bmp image file and save it to a micro-SD card with the filename `testbmp.bmp`. Put this micro-SD card in the SD card reader/writer on the back of the TFT display. Upload the program to the Arduino, and start the Serial Monitor.

The first screen on the TFT display should be a text message indicating that the program has started (Figure 8-4). Next, the program loads a bitmap image from the SD card and displays it on the TFT screen (Figure 8-5). Then a series

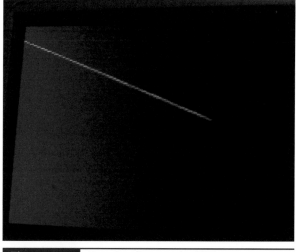

Figure 8-6   Drawing lines to the TFT screen.

of lines is drawn to the TFT screen (Figure 8-6). Next, the program draws a blue or red rectangle to the TFT screen.

On some TFT screens, such as the one I have, the red and blue colors are reversed. For example, the color of the rectangle is set by the `fill()` function, which the official documentation says has the parameters of `fill(red, green, blue)`, with the parameter values ranging from 0 to 255. However, on my version of the TFT screen, the first parameter sets the blue color. This is most likely due to a mistake by the hardware manufacturer. There are many different manufacturers of hardware for the Arduino, so this is not surprising. Thus the following function call sets the rectangle to the color blue on my TFT display, but it is supposed to set the color to red (Figure 8-7):

```
TFTScreen.fill(255,0,0);
```

Now, a green circle is drawn on the TFT screen. The green circle starts off on the left-hand side of the screen and then moves to the right-hand side and then back again (Figure 8-8). Then a series of points is drawn to the screen, and they form a line starting on the left-hand side of the TFT screen and moving to the right-hand side and then moving back (Figure 8-9). Now the screen should indicate that a file is

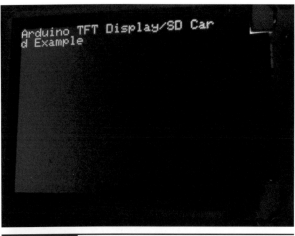

Figure 8-4   The start screen.

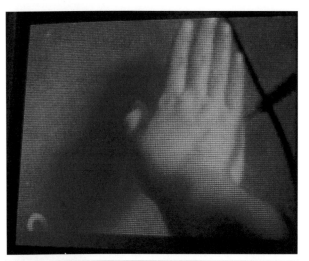

Figure 8-5   Loading and displaying a bitmap image from the SD card.

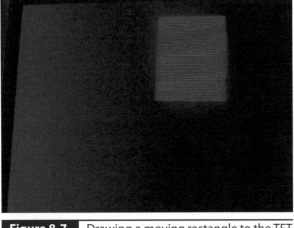

Figure 8-7   Drawing a moving rectangle to the TFT screen.

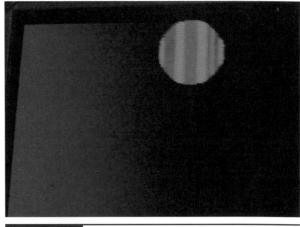

Figure 8-8   Drawing a moving circle to the TFT screen.

being written to the SD card using the SD card reader/writer attached to the back of the TFT display (Figure 8-10). Then the program will read and display the contents of the file from

the SD card that was just written on the TFT screen (Figure 8-11). Finally, the program ends and displays a text message that the program has ended (Figure 8-12).

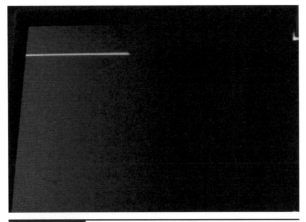

Figure 8-9   Drawing points to form a line.

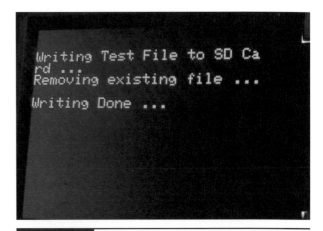

Figure 8-10   Writing a test file to the SD card.

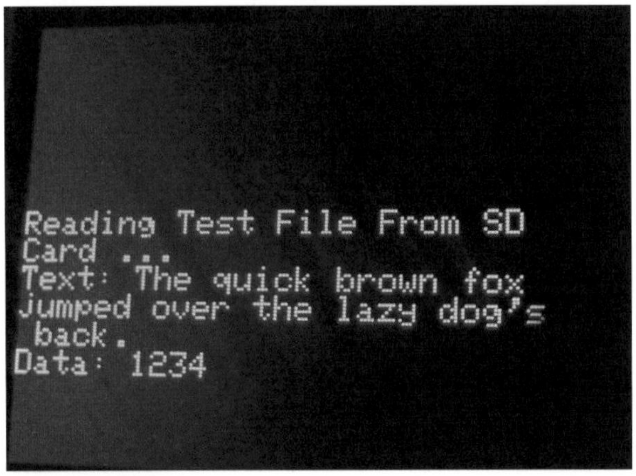

Figure 8-11   Reading the test file from the SD card.

Figure 8-12   The End.

Hands-on Example: Arduino TFT Display Door Entry Alarm with SD Card Logging

This hands-on example project can be easily modified into a full door entry alarm system that detects the opening of a door, logs the number of entries to an SD card, and then displays the status of the alarm on a TFT LCD display. The alarm has a "waiting" state where the alarm cannot be triggered and is designed for a user to be able to set the alarm and then leave the home without the alarm going off. The "waiting" state is followed by the "active" state, where the alarm will detect the presence of a magnet. If a magnet is detected, the alarm state changes to "tripped." The alarm can be reset to the "waiting" state by pressing a button.

Parts List

To build this hands-on example project, you will need

- 1 Arduino TFT LCD display with SD card reader/writer
- 1 micro-SD card
- 1 reed switch
- 1 push button
- 1 10-kΩ resistor
- 1 breadboard (needed to hold the TFT LCD display)
- 1 Arduino Uno (Connection instructions for this hands-on example project are specifically designed for the Arduino Uno.)
- Wires to connect the components to the Arduino

Setting Up the Hardware

To build the hardware for this hands-on example project, you will need to:

- Connect the 5-V pin on the TFT display to the 5-V pin of the Arduino Uno.
- Connect the MISO pin on the TFT display to pin 12 of the Arduino Uno.
- Connect the SCK pin on the TFT display to pin 13 of the Arduino Uno.
- Connect the MOSI pin on the TFT display to pin 11 of the Arduino Uno.
- Connect the LCD CS pin on the TFT display to pin 10 of the Arduino Uno.
- Connect the SD CS pin on the TFT display to pin 4 of the Arduino Uno.
- Connect the D/C pin on the TFT display to pin 9 of the Arduino Uno.
- Connect the RESET pin on the TFT display to pin 8 of the Arduino Uno.
- Connect the BL pin on the TFT display to the 5-V pin of the Arduino Uno.
- Connect the GND pin on the TFT display to the GND pin of the Arduino Uno.
- Connect the VCC pin on the reed switch to the 5-V pin of the Arduino.
- Connect the GND pin on the reed switch to a GND pin of the Arduino.
- Connect the DO pin on the reed switch to pin 7 of the Arduino.
- Connect one terminal on the push button to a node that contains the 10-kΩ resistor and a wire to pin 6 of the Arduino. Connect the other end of the resistor to GND.
- Connect the other terminal on the push button to the 5-V pin of the Arduino (Figure 8-13).

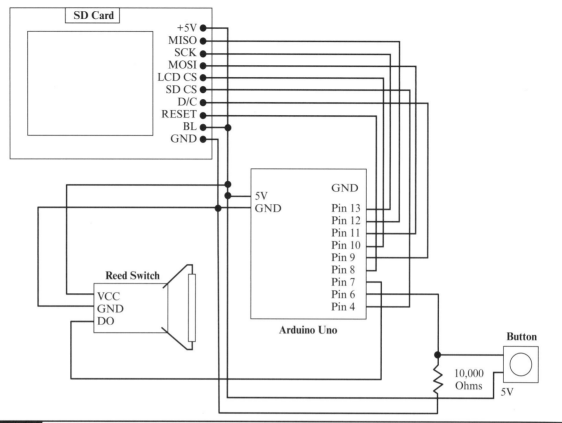

Figure 8-13 Arduino TFT door alarm.

Setting Up the Software

The program for this hands-on example project implements the alarm system by:

1. Including the SPI library so that the TFT display and SD card can be used:

   `#include <SPI.h>`

2. Including the SD card library so that functions needed to read and write data to the SD card can be accessed:

   `#include <SD.h>`

3. Including the TFT library so that functions related to drawing text on the TFT can be used:

   `#include <TFT.h>`

4. Assigning the SD card reader/writer chip select pin to pin 4 of the Arduino:

   `#define sd_cs 4`

5. Assigning the TFT LCD screen chip select pin to pin 10 of the Arduino:

   `#define lcd_cs 10`

6. Assigning the DC pin on the TFT LCD screen to pin 9 of the Arduino:

   `#define dc 9`

7. Assigning the Reset pin on the TFT LCD screen to pin 8 of the Arduino:

   `#define rst 8`

8. Initializing the `TFTScreen` variable, which is a class object and represents the TFT screen, using the `TFT` constructor:

   `TFT TFTScreen = TFT(lcd_cs, dc, rst);`

9. Initializing the `LogFile` variable, which is used to read and write the number of alarm trips to the SD card:

   `File LogFile;`

10. Initializing the `NumberTrips` variable, which holds the total number of times the alarm has been tripped, including past alarm trips from previous power-ups:

    ```
    int NumberTrips = 0;
    ```

11. Assigning the reed sensor output pin to pin 7 of the Arduino:

    ```
    int SensorPin = 7;
    ```

12. Initializing the `RawValue` variable, which holds the input read from devices attached to the Arduino:

    ```
    int RawValue = 0;
    ```

13. Initializing the `AlarmState` enumeration, which holds the different states that the alarm can be in:

    ```
    enum AlarmState
    {
    Waiting,
    Active,
    Tripped
    };
    ```

14. Initializes the `CurrentAlarmState` variable, which holds the current state of the alarm system and is initialized to the `Waiting` state:

    ```
    AlarmState CurrentAlarmState =
      Waiting;
    ```

15. Initializing the `WaitStateTime`, which holds the time in milliseconds that the alarm system stays in the `Waiting` state before it moves to the `Active` state. The default value is 5 seconds.

    ```
    int WaitStateTime = 5000;
    ```

16. Assigning the `ButtonPin` variable, which is used to read in the status of the alarm's reset button, to pin 6 of the Arduino:

    ```
    int ButtonPin = 6;
    ```

The `setup()` function initializes the program by:

1. Setting the `SensorPin` pin of the Arduino (which is connected to the reed switch output) to be an input pin so that data can be read from the reed switch.

2. Setting the `ButtonPin` pin of the Arduino (which is connected to the push button) to be an input pin so that the status of the button can be read. If the voltage is high or 1, the button is being pressed. If the voltage is low or 0, the button is not being pressed.

3. Initializing the Serial Monitor and setting the communication speed to 9,600 baud.

4. Printing a text message indicating that the program has begun to the Serial Monitor.

5. Initializing the TFT LCD display.

6. Clearing the TFT screen and setting the background color to black.

7. Setting the TFT text color to white.

8. Printing a text message to the TFT screen indicating that the program is starting.

9. Suspending execution of the program for 3,000 milliseconds, or 3 seconds.

10. Printing a text message to the Serial Monitor indicating that the SD card reader/writer is being initialized.

11. Initializing the SD card reader/writer.

12. Printing a text message to the Serial Monitor indicating that the SD was not initialized if the SD card was not initialized successfully and then exiting the function.

13. Printing a text message to the Serial Monitor indicating that the SD card reader/writer was initialized successfully.

14. Read the total number of times the alarm has been tripped by calling the `ReadLogFile("Log.txt")` function, which reads the value from a file on the SD card named `Log.txt`.

15. Printing to the TFT screen the value read from step 14.

16. Printing to the Serial Monitor the value read from step 14.

17. Suspending execution of the program for 6 seconds. See Listing 8-12.

Listing 8-12 The `setup()` Function

```
void setup()
{
  pinMode(SensorPin, INPUT);
  pinMode(ButtonPin, INPUT);

  Serial.begin(9600);
  Serial.println(F("TFT Display Door
    Alarm ..."));

  TFTScreen.begin();
  TFTScreen.background(0, 0, 0);
  TFTScreen.stroke(255, 255, 255);
  TFTScreen.println(F("Arduino TFT
   Display Door Alarm ... "));
  delay(3000);

  Serial.print(F("Initializing SD
    card..."));
  if (!SD.begin(sd_cs))
  {
     Serial.println(F("SD Card
       Initialization Failed ..."));
     return;
  }
  Serial.println(F("SD Card
    Initialization OK ..."));

  // Read in and display log file
  ReadLogFile("Log.txt");
  TFTScreen.print("NumberTrips: ");
  TFTScreen.println(String(NumberTrips));

  Serial.print("NumberTrips: ");
  Serial.println(NumberTrips);

  delay(6000);
}
```

The `WriteLogFile()` function writes the total number of times that the alarm has been tripped to a file by:

1. Printing a text message to the Serial Monitor that the log file is about to be written to the SD card.

2. Deleting the log file from the SD card if a file already exists. Printing out the result of the attempt to remove the existing file from the Serial Monitor.

3. Opening the file for writing.

4. If the file was successfully opened for writing, then:

   a. Converting the total number of alarm trips from an integer to a string type and then writing the value to the SD card.

   b. Closing the file.

   c. Printing a text message to the Serial Monitor that the writing of the file has been completed.

5. Printing a text message to the Serial Monitor indicating that opening the file failed if the file was not successfully opened for writing. See Listing 8-13.

The `ReadLogFile()` function reads the number of total times the alarm has been tripped by:

1. Opening the log file on the SD card for reading.

   a. If the log file was successfully opened for reading, while there are more data to read from the file, reading one byte of data and storing it in the character array called `Data` indexed by the variable *i*.

   b. Incrementing the array index variable *i* by 1.

2. Closing the log file.

Listing 8-13 The WriteLogFile() Function

```
void WriteLogFile(char* filename)
{
 Serial.println(F("Writing Number of Alarm Trips to SD Card ..."));
 if (SD.exists(filename))
 {
    if (SD.remove(filename))
    {
       Serial.println(F("Removing existing file ..."));
    }
    else
    {
       Serial.println(F("ERROR ... Failed to remove existing file ..."));
    }
 }

 LogFile = SD.open(filename, FILE_WRITE);
 if (LogFile)
 {
    LogFile.print(String(NumberTrips));
    LogFile.close();
    Serial.println(F("Writing Done ..."));
 }
 else
 {
    Serial.println(F("Error Opening File For Write ..."));
 }
}
```

3. Converting the characters in the character array called Data into an integer by calling the atoi() function and storing the result in the variable NumberTrips. The atoi() function is a standard C library conversion function.

4. Printing a text error message to the Serial Monitor if the log file was not successfully opened for reading. See Listing 8-14.

The ProcessWaitingState() function processes the Waiting state of the alarm by:

1. Clearing the TFT screen and setting the background color of the standard TFT display to red = 0, green = 255, and blue = 255. However, for some TFT screens, the red and blue inputs are switched, which is most likely due to manufacturer error.

2. Setting the color of the text to black.

Listing 8-14 The ReadLogFile() Function

```
void ReadLogFile(char* filename)
{
 char Data[10];
 int i = 0;

 LogFile = SD.open(filename);
 if (LogFile)
 {
    while (LogFile.available())
    {
       Data[i] = LogFile.read();
       i++;
    }
    LogFile.close();
    NumberTrips = atoi(Data);
 }
 else
 {
    Serial.println(F("Error Opening Log
       for Reading ..."));
 }
}
```

3. Displaying text to the TFT screen indicating that the alarm is in the Waiting state.

4. Suspending execution of the program for WaitStateTime milliseconds.

5. Reading the total number of alarm trips from the log file by calling the ReadLogFile("Log.txt") function with the log file name Log.txt.

6. Clearing the TFT screen and setting the background to green.

7. Printing text to the TFT screen indicating that the alarm is now active and printing the total number of times the alarm has been tripped (which was read in from the log file in step 5). See Listing 8-15.

Listing 8-15   The ProcessWaitingState() Function

```
void ProcessWaitingState()
{
 TFTScreen.background(0, 255, 255);
 TFTScreen.stroke(0, 0, 0);
 TFTScreen.text("Alarm Waiting State
   ..", 0, 50);

 delay (WaitStateTime);

 ReadLogFile("Log.txt");
 TFTScreen.background(0, 255, 0);
 TFTScreen.text("Alarm Now Active ...",
   0, 50);
 TFTScreen.println();
 TFTScreen.print("NumberTrips: ");
 TFTScreen.println(String(NumberTrips));
}
```

The loop() function implements the alarm system by repeatedly (in an infinite loop):

1. Reading the status of the alarm reset button.

2. Changing the alarm status to Waiting and printing a text message to the Serial Monitor that the alarm is now in the Waiting state if the alarm reset button is being pressed.

3. If the current alarm state is Waiting, then processing this state by calling the ProcessWaitingState() function. After this function completes, setting the current alarm state to Active.

4. If the current alarm state is Active, then:

    a. Reading the state of the reed switch sensor.

    b. If the reed switch sensor detects a magnet, the alarm has been tripped, so increasing the total number of alarm trips by 1.

    c. Setting the current state of the alarm to Tripped.

    d. Printing the status of the alarm to the Serial Monitor.

    e. Clearing the background of the TFT screen and setting it to red (on my TFT).

    f. Setting the TFT text color to white.

    g. Writing a text message to the TFT screen indicating that the alarm has been tripped.

    h. Writing out a log file that updates the total number of times that the alarm has been tripped by calling the WriteLogFile("Log.txt") function. The Log.txt is the file to write to. See Listing 8-16.

Listing 8-16 The `loop()` Function

```
void loop()
{
 // Read Reset Button
 RawValue = digitalRead(ButtonPin);
 if (RawValue == 1)
 {
    CurrentAlarmState = Waiting;
    Serial.println("CurrentAlarmState:
      Waiting");
 }

 // Check if Alarm is in Waiting State
 if (CurrentAlarmState == Waiting)
 {
    ProcessWaitingState();
    CurrentAlarmState = Active;
 }

 if (CurrentAlarmState == Active)
 {
    RawValue = digitalRead(SensorPin);
    if (RawValue == 0)
    {
      // Alarm Tripped
      NumberTrips++;
      CurrentAlarmState = Tripped;
      Serial.println("CurrentAlarmState:
  Tripped");

      // TFT Update
      TFTScreen.background(0, 0, 255);
      TFTScreen.stroke(255, 255, 255);
      TFTScreen.text("Alarm Tripped
        ...", 0, 50);

      // Write Log File
      WriteLogFile("Log.txt");
    }
 }
}
```

Running the Program

The first thing to note is that on some TFT screens such as mine, the red and blue colors are switched so that when I set the color for blue on my TFT screen, it actually shows up as red. This is most likely a problem with the hardware. Upload the program to your Arduino. You should see the text screen come up on the TFT indicating the program is starting. Next, the alarm should go into the `Waiting` state (Figure 8-14).

Next, the alarm moves into the `Active` state, and the total number of alarm trips is displayed. As you trip and reset the alarm, the number of alarm trips should increase. In Figure 8-15, the alarm is in the `Active` state and has been tripped a total of 22 times.

Figure 8-14 The `Waiting` alarm state shown on the TFT LCD display.

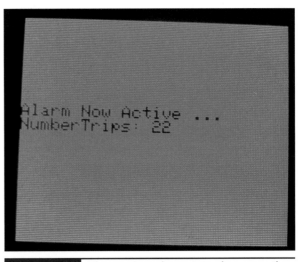

Figure 8-15 The `Active` alarm state shown on the TFT LCD display.

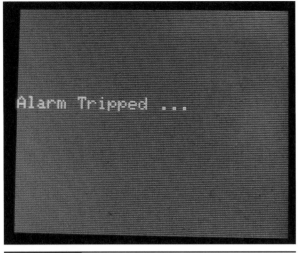

Figure 8-16 The Tripped alarm state shown on the TFT LCD display.

Now bring a magnet close enough to the reed switch that the alarm can be activated. When the alarm is activated, the TFT screen should display the alarm-tripped message (Figure 8-16).

To change this hands-on example project into a working door entry alarm, you would change the alarm to detect the absence of a magnet instead of the presence of a magnet. What you would do is place the reed switch sensor on the door and the magnet on the door frame so that when the door is closed, the magnet would be detected. When the door is open, the magnet would not be detected by the reed switch, which will trigger the alarm.

Arduino Cameras

The two best cameras for the Arduino are the Omnivision OV7670 FIFO camera and the ArduCAM OV2640 minicamera. The OV7670 comes in two main versions, one with first in, first out (FIFO) frame buffer memory and one without any memory. If you are using the camera with the Arduino, then you will need the version with the FIFO memory. The OV7670 is complex to operate and requires an Arduino Mega to use in order to take a picture and save it

on an SD card. The benefit over the ArduCAM mini is that it costs around $9 to $10 dollars, which is less than half the cost of the ArduCAM mini, which is around $25 as of this writing. You can learn more about how to use this camera in a book I wrote called *Beginning Arduino OV7670 Camera Development*, which covers the most popular version, the FIFO version, of the OV7670. The ArduCAM OV2640 costs more than the OV7670 but is much easier to use and only requires an Arduino Uno to take a picture and save it on a SD card. The maximum resolution of the ArduCAM mini is also higher than that of the OV7670. The projects in this book use the ArduCAM OV2640 minicamera.

ArduCAM OV2640 Minicamera

The ArduCAM OV2640 minicamera includes

- 2MP image sensor
- I2C interface for sensor configuration
- SPI interface for camera commands and data stream
- 5-V/3.3-V-tolerant input-output (I/O) ports
- Support for JPEG compression mode, single- and multiple-shoot mode, one-time capture multiple-read operation, burst read operation, low-power mode
- Good mating with standard Arduino boards
- An open source code library for Arduino, STM32, Chipkit, Raspberry Pi, and BeagleBone Black
- Small form factor
- Power supply at 5 V, 70 mA
- Low power mode at 5 V, 20 mA
- Frame buffer of 384 kB
- Resolution support for UXGA, SVGA, VGA, QVGA, CIF, and QCIF
- Image format support for RAW, YUV, RGB, and JPEG (Figure 8-17).

Figure 8-17 ArduCAM OV2640 minicamera.

ArduCAM Minicamera Library Software Installation

To use the ArduCAM minicamera with your Arduino, you will need to download and install the ArduCAM libraries from the ArduCAM website: www.arducam.com. Once you have downloaded the zip file, you will need to uncompress it using a program such as 7-Zip and install the two directories ArduCAM and UTFT4ArduCAM_SPI under the libraries directory for Arduino. For example, on my Windows XP system, I have installed the ArduCAM libraries in my Program Files/ Arduino/libraries directory by copying the two directories to this libraries directory. After doing this, you should be able to compile source code that includes the ArduCAM library.

For the example projects in this book, I used version 3.4.7 of the ArduCAM library, which was released on August 8, 2015.

The Memory Saver Include File

Change the memorysaver.h include file located in the ArduCAM library directory so that the following line is uncommented:

```
#define OV2640_CAM
```

This includes the camera register information needed for the ArduCAM OV2640 mini camera.

Hands-on Example: Arduino ArduCAM OV2640 Mini Portable Programmable Digital Camera System

In this hands-on example project, you will learn how to build a portable programmable camera system using an ArduCAM OV2640 minicamera. You will be able to take a picture by pressing a button that will automatically save the captured image to an SD card. The status of the camera system will be displayed on the TFT screen. This example can serve as a basis for your own custom projects in areas such as security, surveillance, wildlife photography, time-lapse photography, and Internet of things.

Parts List

To build this hands-on example project, you will need

- 1 Arduino TFT LCD display with SD card reader/writer
- 1 micro-SD card
- 1 Arduino Uno
- 1 or more breadboards for the TFT display, Arduino camera, etc.
- 1 push button

- 1 10-kΩ resistor
- 1 ArduCAM OV2640 minicamera

Setting Up the Hardware

To connect the hardware for this hands-on example project, you will need to:

- Connect the 5-V pin on the TFT display to the 5-V pin of the Arduino Uno.
- Connect the MISO pin on the TFT display to pin 12 of the Arduino Uno.
- Connect the SCK pin on the TFT display to pin 13 of the Arduino Uno.
- Connect the MOSI pin on the TFT display to pin 11 of the Arduino Uno.
- Connect the LCD CS pin on the TFT display to pin 10 of the Arduino Uno.
- Connect the SD CS pin on the TFT display to pin 4 of the Arduino Uno.
- Connect the D/C pin on the TFT display to pin 9 of the Arduino Uno.
- Connect the Reset pin on the TFT display to pin 8 of the Arduino Uno.
- Connect the BL pin on the TFT display to the 5-V pin of the Arduino Uno.
- Connect the GND pin on the TFT display to the GND pin of the Arduino Uno.
- Connect the CS pin on the ArduCAM to pin 7 of the Arduino Uno.
- Connect the MOSI pin on the ArduCAM to the node on the breadboard that contains the MOSI pin from the TFT LCD.
- Connect the MISO pin on the ArduCAM to the node on the breadboard that contains the MISO pin from the TFT LCD.
- Connect the SCK pin on the ArduCAM to the node on the breadboard that contains the SCK pin from the TFT LCD.

- Connect the GND pin on the ArduCAM to the GND node or GND pin of the Arduino Uno.
- Connect the VCC pin on the ArduCAM to the 5-V node or 5-V pin of the Arduino Uno.
- Connect the SDA pin on the ArduCAM to analog pin 4 of the Arduino Uno.
- Connect the SCL pin on the ArduCAM to analog pin 5 of the Arduino Uno.
- Connect one terminal on the push button to a node that consists of a 10-kΩ resistor and a wire that connects the node to pin 6 of the Arduino. Connect the other end of the resistor to GND of the Arduino Uno.
- Connect the other terminal on the push button to the 5-V pin of the Arduino Uno (Figure 8-18).

Setting Up the Software

The digital camera system presented in this hands-on example project is controlled by the software discussed in this section. An ArduCAM related library is included:

```
#include <UTFT_SPI.h>
```

The library relating to using the SD card reader/writer is included so that the SD card can be accessed:

```
#include <SD.h>
```

The Wire library is needed for the ArduCAM camera for the I2C interface portion of the camera:

```
#include <Wire.h>
```

The ArduCAM library is included so that the OV2640 minicamera can be used:

```
#include <ArduCAM.h>
```

The SPI library includes code relating to the Serial Peripheral Interface, which is needed

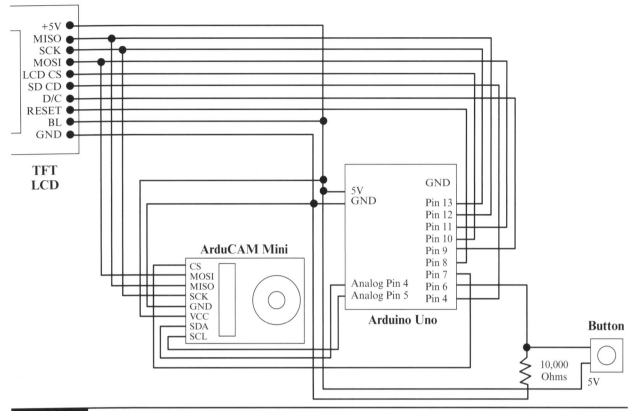

Figure 8-18 Arduino TFT ArduCAM OV2640 minicamera system.

for the TFT display, the SD card, and the ArduCAM camera:

```
#include <SPI.h>
```

The camera's register settings are defined in the memorysaver.h file:

```
#include "memorysaver.h"
```

The TFT library is included so that the TFT display can be initialized and accessed:

```
#include <TFT.h>
```

Set pin 7 of the Arduino as the chip select pin for the ArduCAM minicamera. This pin on the Arduino is connected to the chip select input pin on the camera:

```
const int SPI_CS = 7;
```

Define the ArduCAM minicamera as myCAM, which is of camera type OV2640:

```
ArduCAM myCAM(OV2640, SPI_CS);
```

The ResolutionType enumeration holds the available camera image resolutions. The MAX enumeration means that the camera's resolution is set to 1,280 by 1,024 pixels:

```
enum ResolutionType
{
  MAX,
  VGA,
  QVGA,
  QQVGA
};
```

The Resolution variable holds the current camera resolution. The default is set to VGA resolution. An image that is taken with the camera will be by default in VGA resolution:

```
ResolutionType Resolution = VGA;
```

The chip select pin on the SD card reader/writer is assigned to pin 4 of the Arduino Uno:

```
#define sd_cs 4
```

The chip select pin on the TFT display is assigned to pin 10 of the Arduino Uno:

```
#define lcd_cs 10
```

The dc pin on the TFT display is assigned to pin 9 of the Arduino Uno:

```
#define dc 9
```

The Reset pin on the TFT display is assigned to pin 8 of the Arduino Uno:

```
#define rst 8
```

The TFTScreen variable represents the TFT display and will display the status of the camera system:

```
TFT TFTScreen = TFT(lcd_cs, dc, rst);
```

The DataFile variable is used to write the image that is captured from the ArduCAM to the SD card:

```
File DataFile;
```

The ButtonPin variable is used to get the status of the button and is assigned to pin 6 of the Arduino Uno:

```
int ButtonPin = 6;
```

The status of the button is stored in the RawValue variable:

```
int RawValue = 0;
```

The ImageCount variable holds the number of images taken since the camera system was powered up and is set to the default value of 0:

```
int ImageCount = 0;
```

The ImageFileName is used to create the image filename based on the ImageCount variable:

```
String ImageFileName = "";
```

The FileName variable is a character array that holds the final image filename that will be saved to the SD card:

```
char FileName[10];
```

The OpenFileForWriting() function opens the image file that is to be written to the SD card by:

1. Setting the TFT screen background to black.

2. Setting the text color to white.

3. Printing text to the TFT screen stating that writing the image file to the SD card has started.

4. Printing text to the Serial Monitor stating that writing the image file to the SD card has started.

5. Removing the file and printing a notification text message to the TFT display and Serial Monitor if the file to be written already exists.

6. Printing an error message to the TFT screen and the Serial Monitor if an error occurs when trying to remove the file.

7. Opening the file for writing.

8. Printing a notification message to the TFT display and to the Serial Monitor if the file was opened successfully.

9. Printing an error message to the TFT display and to the Serial Monitor if the file was not opened successfully. See Listing 8-17.

The InitializeCamera() function initializes the ArduCAM OV2640 minicamera for operation by:

1. Initializing the I2C interface by calling the Wire.begin() function if __AVR__ is defined or by calling the Wire1.begin() function if __arm__ is defined.

2. Printing a text message to the Serial Monitor indicating that the ArduCAM camera is initializing.

3. Setting the SPI_CS pin on the Arduino to be an output pin. This pin is connected to the chip select input pin on the ArduCAM mini and is used to determine whether the camera will ignore or process the data coming from the SPI bus.

Listing 8-17 The `OpenFileForWritng()` Function

```
void OpenFileForWriting(char* filename)
{
 TFTScreen.background(0, 0, 0);
 TFTScreen.stroke(255, 255, 255);
 TFTScreen.println(F("Writing File... "));
 Serial.println(F("Writing File... "));

 if (SD.exists(filename))
 {
    if (SD.remove(filename))
    {
       TFTScreen.println(F("Removing file ..."));
       Serial.println(F("Removing file ..."));
    }
    else
    {
       TFTScreen.println(F("ERROR ... Failed to remove existing file ..."));
       Serial.println(F("ERROR ... Failed to remove existing file ..."));
    }
 }

 DataFile = SD.open(filename, FILE_WRITE);
 if (DataFile)
 {
    TFTScreen.print(F("File Opened... "));
    TFTScreen.println(filename);

    Serial.print(F("File Opened... "));
    Serial.println(filename);
 }
 else
 {
    Serial.println(F("Error Opening File For Write ..."));
    TFTScreen.println(F("Error Opening File For Write..."));
 }
}
```

4. Initializing the Serial Peripheral Interface by calling the `SPI.begin()` function.

5. Checking whether the connection between the ArduCAM camera and the Arduino Uno is working correctly by writing `0x55` to the `ARDUCHIP_TEST1` register on the ArduCAM minicamera, reading the register, and verifying that the returned value is `0x55`.

6. If the values do not match, the SPI interface is not working, so printing an error message to the Serial Monitor and suspending execution of the program by entering an infinite loop.

7. Detecting the presence of the OV2640 camera specifically and printing the result to the Serial Monitor.

8. Setting the format of the images to be captured by the camera to JPEG format by calling the myCAM.set_format(JPEG) function with JPEG as the input parameter.

9. Initializing the ArduCAM OV2640 by calling the myCAM.InitCAM() function. See Listing 8-18.

Listing 8-18   The InitializeCamera() Function

```
void InitializeCamera()
{
 uint8_t vid, pid;
 uint8_t temp;
#if defined (__AVR__)
 Wire.begin();
#endif
#if defined(__arm__)
 Wire1.begin();
#endif

 Serial.println(F("ArduCAM Starting .........."));
 // set the SPI_CS as an output:
 pinMode(SPI_CS, OUTPUT);

 // initialize SPI:
 SPI.begin();
 //Check if the ArduCAM SPI bus is OK
 myCAM.write_reg(ARDUCHIP_TEST1, 0x55);
 temp = myCAM.read_reg(ARDUCHIP_TEST1);
 if (temp != 0x55)
 {
    Serial.println(F("SPI interface Error!"));
    while (1);
 }

 //Check if the camera module type is OV2640
 myCAM.rdSensorReg8_8(OV2640_CHIPID_HIGH, &vid);
 myCAM.rdSensorReg8_8(OV2640_CHIPID_LOW, &pid);
 if ((vid != 0x26) || (pid != 0x42))
    Serial.println(F("Can't find OV2640 module!"));
 else
    Serial.println(F("OV2640 detected ..........."));

 // Change to Jpeg capture mode and initialize the OV2640 module
 myCAM.set_format(JPEG);
 myCAM.InitCAM();
}
```

The setup() function initializes the program by:

1. Initializing the Serial Monitor and setting the communication speed to 9,600 baud.

2. Printing a text message to the Serial Monitor indicating that the program has started.

3. Initializing the TFT display.

4. Setting the background color to black.

5. Setting the text color to white.

6. Printing a text message to the TFT display indicating that the program has started.

7. Initializing the camera by calling the InitializeCamera() function.

8. Setting the ButtonPin pin on the Arduino Uno as an input pin so that the status of the button can be read.

9. Printing a text message to the Serial Monitor indicating that the SD card is being initialized.

10. Printing an error message to the Serial Monitor if the SD card reader/writer did not initialize successfully and then exiting the function.

11. Printing a text message to the Serial Monitor indicating that the SD card reader/writer initialized successfully.

12. Printing text messages to the TFT display and the Serial Monitor that the camera system is ready to take photos. See Listing 8-19.

Listing 8-19 The setup() Function

```
void setup()
{
 Serial.begin(9600);
 Serial.println(F("TFT ArduCAM Test ..."));

 TFTScreen.begin();
 TFTScreen.background(0, 0, 0);
 TFTScreen.stroke(255, 255, 255);
 TFTScreen.println(F("TFT ArduCAM Test ... "));

 // Initializes the ArduCAM Mini Camera
 InitializeCamera();

 // Initialize ButtonPin to read in Button Status
 pinMode(ButtonPin, INPUT);

 // SD Card
 Serial.print(F("Initializing SD card..."));
 if (!SD.begin(sd_cs))
 {
    Serial.println(F("SD Card Initialization Failed ..."));
    return;
 }
 Serial.println(F("SD Card Initialization OK ..."));

 TFTScreen.println(F("Ready To Take Photo..."));
 Serial.println(F("Ready To Take Photo..."));
 }
```

The CaptureImage() function captures an image to the ArduCAM minicamera's frame buffer memory by:

1. Printing a text message to the Serial Monitor indicating that an image is being captured by the camera.

2. Resetting the FIFO frame buffer read pointer to 0 by calling the myCAM.flush_fifo() function. The frame buffer holds the captured image.

3. Clearing the flag that is automatically set after an image is captured by calling the myCAM.clear_fifo_flag() function.

4. Capturing an image by calling the myCAM. start_capture() function. This function will wait until the start of a new frame and then capture that frame to the FIFO frame buffer memory on the camera.

5. Printing a text message to the Serial Monitor indicating that the camera has started to capture an image.

6. Polling the camera continuously until the capture done flag has been set, indicating that the entire image has just finished being captured by the camera.

7. Printing a text message to the Serial Monitor indicating that the image has just finished being captured by the camera. See Listing 8-20.

The ReadFifoBurstWriteSDCard() function reads the image from the camera's FIFO memory frame buffer and saves the image to a SD card using the SD card reader/writer attached to the back of the TFT display by:

1. Getting the length of the captured image in bytes by calling the myCAM.read_fifo_ length() function.

2. Exiting the function if the length of the captured image is greater than or equal to the size of the FIFO frame buffer, which is 384 kB. This means that an error occurred while capturing the image, and the image was not found.

Listing 8-20 The CaptureImage() Function

```
void CaptureImage()
{
// Capture Image from camera to camera's FIFO memory buffer
Serial.println(F("... Starting Image Capture ..."));

//Flush the FIFO
myCAM.flush_fifo();

//Clear the capture done flag
myCAM.clear_fifo_flag();

//Start capture
myCAM.start_capture();
Serial.println(F("Start Capture"));

// Wait till capture is finished
while (!myCAM.get_bit(ARDUCHIP_TRIG , CAP_DONE_MASK));
Serial.println(F("Capture Done!"));
}
```

3. Printing the length of the captured image to the Serial Monitor.

4. Getting the current system time since power-up in milliseconds by calling the `millis()` function.

5. Setting the camera to recognize data from the Serial Peripherals Interface by calling the `myCAM.CS_LOW()` function. This function sets the voltage on the chip select pin on the camera to 0 or LOW.

6. Setting the FIFO memory to be read in burst mode by calling the `myCAM.set_fifo_burst()` function.

7. Reading in the dummy byte from the camera's FIFO buffer to the Arduino Uno by calling the `SPI.transfer(0x00)` function.

8. Decreasing the counter that keeps track of the number of bytes needed to be read from the frame buffer by 1.

9. While there are still more bytes to read from the camera's buffer, then:

   a. Reading a byte from the FIFO frame buffer on the camera by calling the `SPI.transfer(0x00)` function and storing the result in the variable `temp`.

   b. Disabling the camera from SPI operations by calling the `myCAM.CS_HIGH()` function, which sets the chip select pin on the camera to 1 or HIGH voltage.

   c. Writing the byte to the SD card by calling the `DataFile.write(temp)` function with `temp` as the input parameter. Remember that `temp` holds the data that were just read from the camera.

   d. Enabling the camera to use the SPI by calling the `myCAM.CS_LOW()` function. This sets the chip select pin on the camera to 0 or LOW voltage.

   e. Setting the camera's memory into FIFO burst mode by calling the `myCAM.set_fifo_burst()` function.

   f. Increasing the `bytecount` variable by 1. This variable keeps track of the total number of bytes read from the camera's memory.

   g. Suspending execution of the program by 10 microseconds by calling the `delayMicroseconds(10)` function with input parameter 10.

10. Disabling the camera from using the SPI by calling the `myCAM.CS_HIGH()` function, which sets the chip select pin on the camera to 1 or HIGH voltage.

11. Closing the image file that was just written by calling the `DataFile.close()` function.

12. Printing the total number of bytes written to the file to the Serial Monitor.

13. Printing the total time in milliseconds that it took to write the image file to the SD card. See Listing 8-21.

Listing 8-21 The `ReadFifoBurstWriteSDCard()` Function

```
uint8_t ReadFifoBurstWriteSDCard()
{
 uint32_t length = 0;
 uint8_t temp,temp_last;

 long bytecount = 0;
```

(continued on next page)

Listing 8-21   The `ReadFifoBurstWriteSDCard()` Function (*continued*)

```
int total_time = 0;

length = myCAM.read_fifo_length();
if(length >= 393216) //384 kb
{
   Serial.println(F("Not found the end."));
   return 0;
}
Serial.print(F("READ_FIFO_LENGTH() = "));
Serial.println(length);

total_time = millis();
myCAM.CS_LOW();
myCAM.set_fifo_burst();
temp = SPI.transfer(0x00); // Added in to original source code, read in dummy byte

length--;
while( length-- )
{
    temp = SPI.transfer(0x00);

   //////// Write byte to SD Card
   // Deactivate camera
   myCAM.CS_HIGH();

   // Write Byte to file
   DataFile.write(temp);

   // Activate Camera
   myCAM.CS_LOW();
   myCAM.set_fifo_burst();
   /////////////////////////////

   bytecount++;
   delayMicroseconds(10);
}
myCAM.CS_HIGH();

// Close File on SD Card
DataFile.close();

Serial.print(F("ByteCount = "));
Serial.println(bytecount);

total_time = millis() - total_time;
Serial.print(F("Total time used:"));
Serial.print(total_time, DEC);
Serial.println(F(" milliseconds ..."));
}
```

The `SetCameraResolution()` function sets the camera resolution based on the `Resolution` variable by:

1. Changing the camera resolution to 160 by 120 pixels if `Resolution` is set to QQVGA.

2. Changing the camera resolution to 320 by 240 pixels if `Resolution` is set to QVGA.

3. Changing the camera resolution to 640 by 480 pixels if `Resolution` is set to VGA.

4. Changing the camera resolution to 1,280 by 1,024 pixels if `Resolution` is set to MAX.

5. Printing an error message to the Serial Monitor if `Resolution` is none of the preceding values. See Listing 8-22.

The `loop()` function contains the main program code for this camera system and in a continuous loop:

1. Reads the status of the push button.

2. If the button is pressed, then:

   a. Prints a text message to the Serial Monitor indicating that the camera is now taking a photo.

   b. Sets the background color of the TFT screen to red on my TFT or blue on a standard TFT display.

   c. Sets the text color to white.

   d. Displays text on the TFT screen indicating that the camera is now taking a photo.

Listing 8-22 The `SetCameraResolution()` Function

```
void SetCameraResolution()
{
  // Set Screen Size
  if (Resolution == QQVGA)
  {
     myCAM.OV2640_set_JPEG_size(OV2640_160x120);
  }
  else
  if (Resolution == QVGA)
  {
     myCAM.OV2640_set_JPEG_size(OV2640_320x240);
  }
  else
  if (Resolution == VGA)
  {
     myCAM.OV2640_set_JPEG_size(OV2640_640x480);
  }
  else
  if (Resolution == MAX)
  {
     myCAM.OV2640_set_JPEG_size(OV2640_1280x1024);
  }
  else
  {
     Serial.println(F("ERROR in setting Camera Resolution"));
  }
}
```

e. Sets the camera resolution to VGA by setting the `Resolution` variable to VGA and then calling the `SetCameraResolution()` function.

f. Captures an image using the camera by calling the `CaptureImage()` function.

g. Increases the number of photos taken by 1.

h. Creates the filename for the photo by concatenating the number of photos taken and the .jpg file extension.

i. Creates and opens a file with the filename from step h for saving the image by calling the `OpenFileForWriting(FileName)` function.

j. Reads the image data from the camera's FIFO frame buffer and saves this image data to the file that was opened in step i by calling the `ReadFifoBurstWriteSDCard()` function.

k. Prints text to the TFT screen indicating that the image file has been saved to the SD card. See Listing 8-23.

Listing 8-23 The `loop()` Function

```
void loop()
{
 // Read Button Status
 RawValue = digitalRead(ButtonPin);
 if (RawValue == 1)
 {
    // Display Taking Photo Message
    Serial.println(F("Taking Photo ... "));
    TFTScreen.background(0, 0, 255);
    TFTScreen.stroke(255, 255, 255);
    TFTScreen.text("Taking Photo ...",0,60);
    TFTScreen.println();

    // Set Camera's Resolution
    Resolution = VGA;
    SetCameraResolution();

    // Capture Image to ArduCAM's FIFO Memory
    CaptureImage();

    // Write Image File to SD Card
    ImageCount++;
    ImageFileName = String(ImageCount) + ".jpg";
    strcpy(FileName,ImageFileName.c_str());
    OpenFileForWriting(FileName);
    ReadFifoBurstWriteSDCard();

    TFTScreen.println();
    TFTScreen.println(F("Finished Saving Photo ..."));
 }
}
```

Running the Program

Upload the program to your Arduino, and start the Serial Monitor. The startup message indicating that the program has begun should be printed to the Serial Monitor:

```
TFT ArduCAM Test ...
```

The ArduCAM should be initializing, and the OV2640 mini camera should be detected:

```
ArduCAM Starting ..........
OV2640 detected ...........
```

Next, the SD card is initialized, and a message should be printed to the Serial Monitor indicating that the SD has been initialized successfully:

```
Initializing SD card...SD Card
Initialization OK ...
```

In addition, a message should be displayed on the Serial Monitor and the TFT screen indicating that the camera system is ready to take a photo and save it to the SD card:

```
Ready To Take Photo...
```

To take a photo, press the button. You should see messages on the Serial Monitor indicating that a photo is being taken by the camera:

```
Taking Photo ...
... Starting Image Capture ...
Start Capture
```

When the capture is completed, the following message should appear:

```
Capture Done!
```

Now the image is written to the SD card. If a file with the same filename already exists, then that file is deleted and a new image file is written. The length of the image file in the FIFO memory, the final size of the file that is written, and the total time to write the file are also displayed on the Serial Monitor:

```
Writing File...
Removing file ...
File Opened... 1.jpg
READ_FIFO_LENGTH() = 9216
ByteCount = 9215
Total time used:615 milliseconds ...
```

This file should now be saved on the SD card. Turn off the Arduino, and remove the SD card. Read and display the image file on the SD card using your computer. An example image from the camera I took is shown in Figure 8-19. Another image I took using the camera system is shown in Figure 8-20.

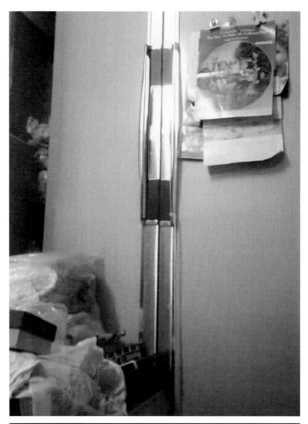

Figure 8-19 Image taken with the ArduCAM OV2640 camera system.

Figure 8-20 Image taken with the ArduCAM OV2640 camera system.

Summary

This chapter has covered the TFT display with SD card reader/writer and the ArduCAM OV2640 minicamera. First, we discussed the TFT display, and then we provided a hands-on example project demonstrating the features of the TFT display and the attached SD card reader/writer that is located on the back of the display. You learned how to display text and display and animate lines, rectangles, circles, and points using the display. You also learned how to read and write files using the SD card reader/writer. Next, a hands-on example project used the TFT display, the SD card, and a reed switch to create a door entry alarm system. Then we discussed the ArduCAM OV2640 minicamera. This was followed by a hands-on example where this camera was used as part of a larger camera system by which you could take a picture and save it to an SD card by pressing a button.

Index